This series aims to publish some of the most relevant results presented at the ICIAM 2023 held in Tokyo in August 2023.

The series is managed by an independent Editorial Board, and will include peer-reviewed content only, including the Invited Speakers volume as well as books resulting from mini-symposia and collateral workshops.

The series is aimed at providing useful reference material to academic and researchers at an international level.

Jiro Akahori • Syoiti Ninomiya • Toshihiro Yamada
Editors

Recent Developments in Stochastic Numerics and Computational Finance

Springer

Editors
Jiro Akahori
Department of Mathematical Sciences
Ritsumeikan University
Kusatsu, Japan

Syoiti Ninomiya
Department of Mathematics, School of Science
Institute of Science Tokyo
Tokyo, Japan

Toshihiro Yamada
Graduate School of Economics
Hitotsubashi University
Tokyo, Japan

ICIAM international congress on industrial and applied mathematics international congress on industrial and applied mathematics2023 ICIAM Tokyo, Japan 2023 08 19 2023 08 24 https://iciam2023.org/

ISSN 3091-3098 ISSN 3091-3101 (electronic)
ICIAM2023 Springer Series
ISBN 978-981-95-0651-4 ISBN 978-981-95-0652-1 (eBook)
https://doi.org/10.1007/978-981-95-0652-1

Mathematics Subject Classification: 60H35, 65C30, 91G15, 91G30, 91G10, 91G20

This Springer imprint is published by the registered company Springer Nature Singapore Pte Ltd.
The registered company address is: 152 Beach Road, #21-01/04 Gateway East, Singapore 189721, Singapore

Preface

Stochastic numerical analysis plays an important role in probability theory, statistics, and applied mathematics, especially in the fields of machine learning and data science, and has achieved a great success in computational finance. In August 2023, "Recent Developments in Stochastic Numerics and Computational Finance" was held as a symposium in ICIAM 2023 Tokyo (10th International Congress on Industrial and Applied Mathematics) organized by Jiro Akahori, Syoiti Ninomiya, and Toshihiro Yamada in order to review recent and new methods for solving various problems on stochastic numerical analysis and computational finance. With many participants from Asia and Europe, the symposium was a big success. We would like to acknowledge all speakers and participants of the symposium.

The aim of this book is to highlight recent advances in stochastic numerical analysis and computational finance. This volume is a part of a series dedicated to ICIAM 2023 Tokyo, and the papers appearing in the chapters of the volume are authored by some of the speakers of the symposium.

The contents will cover a wide range of topics on numerical methods of stochastic differential equations, partial differential equations, and stochastic partial differential equations for pricing derivatives and portfolio optimization, and their theoretical analysis. The book provides new computational schemes and algorithms, including machine learning and deep learning technique for various financial models in computational finance, which are important both in theory and practice.

We believe that the volume will invite graduate students, researchers, and practitioners who are interested in stochastic numerical analysis, financial mathematics, and mathematical finance to find new topics and methods in computational finance.

Kusatsu, Japan
Tokyo, Japan
Tokyo, Japan
January 2025

Jiro Akahori
Syoiti Ninomiya
Toshihiro Yamada

Contents

Policy Improvement Algorithm for an Optimal Consumption and Investment Problem Under a Certain Nonlinear Stochastic Factor Model

Hiroaki Hata and Kazuhiro Yasuda

Abstract Recently, Hata et al. (SIAM J. Cont. Optim. **56**, 3149–3183, 2018) employed a general nonlinear factor model and solved for a consumption and investment problem on a finite time horizon to optimize a discounted expected power utility of consumption and terminal wealth. Further, Hata and Yasuda (Policy improvement algorithm for an optimal consumption and investment problem under general stochastic factor models (2024, preprint)) proposed a policy improvement algorithm, which establishes an iteration procedure converging to the value function and the optimal strategies obtained in Hata et al. (SIAM J. Cont. Optim. **56**, 3149–3183, 2018). Here, we employ a certain nonlinear stochastic factor model and provide some numerical results to understand convergence behaviors of the algorithm.

Keywords Merton problem · Policy improvement algorithm · Nonlinear stochastic factor model · Incomplete market

Mathematics Subject Classification (2020) 93E20, 65M12, 91B28, 60H30, 49L20

1 Introduction

Hata et al. [1] addressed an optimal consumption and investment problem over a finite time horizon to maximize the discounted expected power utility of consumption and terminal wealth. It is well known that this problem, under the

H. Hata
Graduate School of Business Administration, Hitotsubashi University, Kunitachi, Tokyo, Japan
e-mail: h.hata@r.hit-u.ac.jp

K. Yasuda (✉)
Faculty of Science and Engineering, Hosei University, Koganei, Tokyo, Japan
e-mail: k_yasuda@hosei.ac.jp

J. Akahori et al. (eds.), *Recent Developments in Stochastic Numerics and Computational Finance*, ICIAM2023 Springer Series 6,
https://doi.org/10.1007/978-981-95-0652-1_1

Black-Scholes model, was first proposed by Merton [5]. Since then, this problem has been extensively studied by numerous researchers and practitioners. Hata et al. [1] tackled the problem in the case of a general nonlinear stochastic factor model in an incomplete market. However, in [1], the optimal consumption-investment strategy was not explicitly formulated, and no explicit optimal value was obtained. Therefore, Hata and Yasuda [2] formulated a policy improvement algorithm (PIA), which is a sequence of approximations for the optimal strategy and optimal value as determined in [1]. In particular, the solution at a certain step can be expressed by utilizing the solution from the previous step along with a stochastic representation. This algorithm was originally proposed by Howard [4] to solve Markovian dynamic programming problems on a finite state space. Subsequently, this algorithm has been applied to more complex problems. Hata and Yasuda [2] provides some numerical illustrations about convergences of PIA for a linear stochastic factor model. In this paper, we implement the PIA formulated in [2], using a nonlinear stochastic factor model where the mean return rate of the risky asset and the drift of the factor process are expressed as a linear function of the arctangent of the factor process.

The paper is organized as follows. In Sect. 2, we introduce our model and recall results of the original optimal consumption and investment problem obtained in [1]. In Sect. 3, we recall the PIA formulated in [2]. In Sect. 4, we show some numerical results to illustrate our convergence.

2 An Optimal Consumption and Investment Problem

In this section, we first recall the optimal consumption and investment problem studied by Hata et al. [1]. We then give the assumption used in this paper. We introduce the results of the HJB equation derived by applying the dynamic programming approach. And, we also introduce the verification theorem that the candidate of the optimal strategy obtained from the HJB equation is truly optimal.

2.1 Stochastic Factor Model

Consider a market consisting of one bank account and m risky stocks. Suppose that the bank account process S^0 and the price processes of the risky stocks $S := (S^1, \cdots, S^m)^*$ are governed by the equations:

$$\begin{aligned} dS_t^0 &= r(Y_t)S_t^0 dt, \quad S_0^0 = s_0^0, \\ dS_t^i &= S_t^i \left\{ \mu^i(Y_t)dt + \sum_{k=1}^{m+n} \sigma_k^i(Y_t)dW_t^k \right\}, \quad S_0^i = s_0^i, \\ dY_t &= g(Y_t)dt + \lambda(Y_t)dW_t, \quad Y_0 = y \in \mathbb{R}^n, \end{aligned} \tag{1}$$

where $W_t = (W_t^k)_{k=1,\cdots,m+n}$ is an $m+n$ dimensional standard Brownian motion defined on a filtered probability space $(\Omega, \mathcal{F}, P, \{\mathcal{F}_t\})$. We can take $\{\mathcal{F}_t\} = \{\mathcal{F}_t^W\}$, the filtration generated by the Brownian motion W. We denote $(\cdot)^*$ the transpose of a vector or a matrix. Here Y is the n-dimensional stochastic factor process which affects the risk-free interest rate $r(Y_t)$, the mean return drift $\mu(Y_t)$ of S, and the volatility matrix $\sigma(Y_t)$ of S. σ and λ are $m \times (m+n)$, $n \times (m+n)$ matrix valued functions respectively and μ and g are $\mathbb{R}^m$-valued, $\mathbb{R}^n$-valued functions respectively.

We always assume the following conditions:

(H1) λ, g, σ, μ and r are globally Lipschitz and smooth.
(H2) There are constants $\mu_1, \mu_2 > 0$ such that for $x, \eta \in \mathbb{R}^n, \xi \in \mathbb{R}^m$,

$$\mu_1|\xi|^2 \leq \xi^*\sigma(x)\sigma(x)^*\xi \leq \mu_2|\xi|^2, \quad \mu_1|\eta|^2 \leq \eta^*\lambda(x)\lambda(x)^*\eta \leq \mu_2|\eta|^2.$$

(H3) r is nonnegative and bounded with bounded derivatives.

2.2 *Formulation of Optimal Consumption and Investment Problem*

Consider an investor who invests at time t a proportion π_t^i of the investor's wealth in the ith risky asset with price S_t^i, $i = 1, \ldots, m$. With $\pi_t = (\pi_t^1, \ldots, \pi_t^m)^*$ chosen, the proportion of wealth invested in the bank account is $1 - \pi_t^*\mathbf{1}$. Here $\mathbf{1} = (1, \ldots, 1)_m^*$, where $(1, \ldots, 1)_m^*$ is the vector in $\mathbb{R}^m$ with component 1 everywhere. Furthermore, if we consider X_t as the investor's wealth, then $c_t X_t$ becomes the consumption rate.

By the self-financing condition, the investor's wealth $X_t = X_t^{c,\pi}$, starting with the initial capital x solves the following :

$$\begin{aligned}\frac{dX_t^{c,\pi}}{X_t^{c,\pi}} &= \left(1 - \sum_{i=1}^{m} \pi_t^i\right)\frac{dS_t^0}{S_t^0} + \sum_{i=1}^{m} \pi_t^i \frac{dS_t^i}{S_t^i} - c_t dt \\ &= \left\{r(Y_t) + \pi_t^*(\mu(Y_t) - r(Y_t)\mathbf{1}) - c_t\right\}dt + \pi_t^*\sigma(Y_t)dW_t.\end{aligned} \tag{2}$$

with $X_0^{c,\pi} = x$. In this paper, we aim to maximize the discounted expected power utility of consumption and terminal wealth on a finite time horizon : for given constants $T > 0$ and $\gamma \in (-\infty, 0)$, we consider

$$J(T, x, y, c, \pi) := E\left[\int_0^T e^{-\rho t}\frac{1}{\gamma}(c_t X_t^{c,\pi})^\gamma dt + e^{-\rho T}\frac{1}{\gamma}(X_T^{c,\pi})^\gamma\right], \quad \rho > 0. \tag{3}$$

In other words, we tackle the problem of maximizing (3):

$$V(0,x,y) := \sup_{(c,\pi)\in\mathcal{A}_T} J(T,x,y,c,\pi), \tag{4}$$

where $\mathcal{A}_T(\subset \mathcal{L}_{1,2,T})$ is the set of admissible strategies defined in (18). $\mathcal{L}_{1,2,T}$ is defined by

$$\begin{aligned}\mathcal{L}_{1,2,T} := \Big\{(c_t,\pi_t)_{t\in[0,T]};\ &(c_t,\pi_t) \text{ is a } [0,\infty)\\ &\times\mathbb{R}^m\text{-valued } \mathcal{F}_t\text{-progressively measurable,}\\ &\text{stochastic process such that } \int_0^T c_t dt < \infty,\\ &\text{and } \int_0^T |\pi_t|^2 dt < \infty,\ P-a.s.\Big\}.\end{aligned}$$

Using (2) and (3), we have

$$\begin{aligned}J(T,x,y,c,\pi) = \frac{x^\gamma}{\gamma}E\Bigg[&\int_0^T c_t^\gamma e^{\int_0^t [\gamma\{\ell(Y_u,\pi_u)-c_u\}-\rho]du}\tilde{M}_t^\pi dt\\ &+e^{\int_0^T [\gamma\{\ell(Y_t,\pi_t)-c_t\}-\rho]dt}\tilde{M}_T^\pi\Bigg],\end{aligned} \tag{5}$$

where $\ell(y,\pi)$ and $\tilde{M}_T^\pi$ are defined as follows:

$$\ell(y,\pi) := -\frac{1-\gamma}{2}\pi^*\sigma(y)\sigma(y)^*\pi + \pi^*(\mu(y)-r(y)\mathbf{1}) + r(y), \tag{6}$$

$$\tilde{M}_t^\pi := e^{\gamma\int_0^t \pi_s^*\sigma(Y_s)dW_s - \frac{\gamma^2}{2}\int_0^t \pi_s^*\sigma(Y_s)\sigma(Y_s)^*\pi_s ds}. \tag{7}$$

Define $\mathcal{A}_T^1 := \Big\{(c_t,\pi_t)\in\mathcal{L}_{1,2,T};\ E[\tilde{M}_T^\pi]=1\Big\}$. Then, for all $(c,\pi)\in\mathcal{A}_T^1$ we define the probability measure $\tilde{P}^\pi$ on $\mathcal{F}_T$ given by $\left.\frac{d\tilde{P}^\pi}{dP}\right|_{\mathcal{F}_T} := \tilde{M}_T^\pi$. Then, $\tilde{W}_t^\pi := W_t - \gamma\int_0^t \sigma(Y_u)^*\pi_u du$ is a $\tilde{P}^\pi$-Brownian motion, Y_t under $\tilde{P}^\pi$, solves

$$dY_t = \big\{g(Y_t)+\gamma\lambda(Y_t)\sigma(Y_t)^*\pi_t\big\}dt + \lambda(Y_t)d\tilde{W}_t^\pi, \quad Y_0 = y. \tag{8}$$

So, we have, for $(c,\pi)\in\mathcal{A}_T^1$,

$$J(T,x,y,c,\pi) = \frac{x^\gamma}{\gamma}\widetilde{J}(T,y,c,\pi).$$

Here, $\widetilde{J}(T, y, c, \pi)$ is defined by

$$\widetilde{J}(T, y, c, \pi) := \tilde{E}^{\pi}\left[\int_0^T e^{-\rho t} c_t^{\gamma} e^{\gamma \int_0^t \{\ell(Y_u,\pi_u)-c_u\}du} dt + e^{-\rho T} e^{\gamma \int_0^T \{\ell(Y_t,\pi_t)-c_t\}dt}\right], \tag{9}$$

where $\tilde{E}^{\pi}[\cdot]$ is the expectation with respect to $\tilde{P}^{\pi}$.

Now we consider, for $0 \le t \le T < \infty$,

$$v(t, y) := \inf_{(c,\pi)\in\mathcal{A}^1_{T-t}} \widetilde{J}(T-t, y, c, \pi). \tag{10}$$

The dynamic programming approach leads to the derivation of the following Hamilton-Jacobi-Bellman (HJB) equation for (10) with dynamics (8):

$$\frac{\partial v}{\partial t} + \frac{1}{2}\mathrm{tr}(\lambda(y)\lambda(y)^* D^2 v) + g(y)^* Dv - \rho v + \inf_{c\ge 0}\left(-\gamma c v + c^{\gamma}\right) + \inf_{\pi\in\mathbb{R}^m}\left[\gamma\left\{\ell(y,\pi)v + \pi^*\sigma(y)\lambda(y)^* Dv\right\}\right] = 0,$$

where $v(T, y) = 1$. Here $Dv = (D_{y_1}v, \cdots, D_{y_n}v)^*$ and $D^2 v = (D_{y_i y_j}v)$. In the following, we will also use $D_{ij}v$ for $D_{y_i y_j}v$. Assuming that

$$v(t, y) := e^{Z(t,y)},$$

it follows that Z fulfills the subsequent partial differential equation:

$$\frac{\partial Z}{\partial t} + \frac{1}{2}\mathrm{tr}(\lambda(y)\lambda(y)^* D^2 Z) + \frac{1}{2}(DZ)^*\lambda(y)\lambda(y)^* DZ + g(y)^* DZ - \rho + \inf_{c\ge 0}\left(-\gamma c + c^{\gamma} e^{-Z}\right) + \inf_{\pi\in\mathbb{R}^m}\left[\gamma\left\{\ell(y,\pi) + \pi^*\sigma(y)\lambda(y)^* DZ\right\}\right] = 0, \tag{11}$$

with $Z(T, y) = 0$. Observe that $(c^Z(t, y), \pi^Z(t, y))$ defined by

$$c^Z(t, y) := e^{-\frac{Z(t,y)}{1-\gamma}},$$

$$\pi^Z(t, y) := \frac{1}{1-\gamma}(\sigma(y)\sigma(y)^*)^{-1}\left\{\mu(y) - r(y)\mathbf{1} + \sigma(y)\lambda(y)^* DZ(t, y)\right\}.$$

is the minimizer. Then, (11) turns into

$$\frac{\partial Z}{\partial t} + \frac{1}{2}\mathrm{tr}(a(y)D^2 Z) + \frac{1}{2}(DZ)^* Q(y) DZ + (1-\gamma)e^{-\frac{Z}{1-\gamma}} + B(y)^* DZ - U(y) - \rho = 0, \quad Z(T, y) = 0, \tag{12}$$

where a, Q, B and U are defined as follows:

$$
\begin{aligned}
a(y) &:= \lambda(y)\lambda(y)^*, \\
Q(y) &:= \lambda(y)(I + \frac{\gamma}{1-\gamma}\sigma(y)^*(\sigma(y)\sigma(y)^*)^{-1}\sigma(y))\lambda(y)^*, \\
B(y) &:= g(y) + \frac{\gamma}{1-\gamma}\lambda(y)\sigma(y)^*(\sigma(y)\sigma(y)^*)^{-1}(\mu(y) - r(y)\mathbf{1}), \\
U(y) &:= -\frac{\gamma}{2(1-\gamma)}(\mu(y) - r(y)\mathbf{1})^*(\sigma(y)\sigma(y)^*)^{-1}(\mu(y) - r(y)\mathbf{1}) - \gamma r(y).
\end{aligned}
\tag{13}
$$

Here, I is the $(m+n) \times (m+n)$ identity matrix.

2.3 Existence and Uniqueness of Solutions for HJB Equation

In this subsection, we cite the results of the existence and uniqueness of the solution to the HJB equation, as proven in [1].

Set

$$
\hat{Z}(t, y) = (1-\gamma)\log E_{t,y}\left[\int_t^T \mathrm{e}^{-\int_t^s \frac{1}{1-\gamma}\{U(\bar{Y}_u)+\rho\}du} ds + \mathrm{e}^{-\int_t^T \frac{1}{1-\gamma}\{U(\bar{Y}_u)+\rho\}du}\right],
\tag{14}
$$

$$
\bar{Z}(t, y) = \log E_{t,y}\left[\int_t^T \mathrm{e}^{-\int_t^s \{U(\bar{Y}_u)+\gamma+\rho\}du} ds + \mathrm{e}^{-\int_t^T \{U(\bar{Y}_u)+\gamma+\rho\}du}\right].
$$

where $\bar{Y}_s$ solves the following stochastic differential equation:

$$
d\bar{Y}_s = B(\bar{Y}_s)ds + \lambda(\bar{Y}_s)dW_s, \quad \bar{Y}_t = y, \quad s \in [t, T].
$$

Then, we directly cite the results of the existence and uniqueness of the solution to the HJB equation (12) obtained in [1].

Theorem 1 ([1] : Theorems 3.8 and 3.9) *Assume **(H1)** $\sim$ **(H3)**.*

1. *Equation* (12) *has a solution* $\tilde{Z} \in C^{1+\frac{\alpha}{2},2+\alpha}((0,T) \times \mathbb{R}^n) \cap C([0,T] \times \mathbb{R}^n)$. *Further, it satisfies the following estimates:*

$$
\hat{Z} \leq \tilde{Z} \leq \bar{Z}, \tag{15}
$$

$$
\frac{\partial \tilde{Z}}{\partial t} + 1 - \gamma \geq 0, \tag{16}
$$

$$|D\tilde{Z}(t,y)|^2 + \theta\left\{\frac{\partial \tilde{Z}}{\partial t}(t,y) + 1 - \gamma\right\} \leq C\left\{|Da|_{2r}^2 + |Q|_{2r}^2 \right. \tag{17}$$

$$\left. + |DQ|_{2r}^2 + |B|_{2r}^2 + |DB|_{2r} + |U|_{2r} + |DU|_{2r} + 1\right\}, \ y \in B_r, t \in [0,T],$$

where C is a positive constant independent of r and T, θ is a sufficiently large constant, and $|\cdot|_{2r} := ||\cdot||_{L^\infty(B_{2r})}$. Here, $B_r := \{y \in \mathbb{R}^n; |y| \leq r\}$.

2. *Assume Z_0 to be a solution of* (12) *satisfying the conditions in Theorem* 1. *Then, for any solution Z of* (12), *we have*

$$Z \geq Z_0, \ \textit{on } [0,T] \times \mathbb{R}^n,$$

which implies that the solution satisfying (15), (16) *and* (17) *is unique.*

If a solution Z satisfies the condition that there is $C > 0$ such that

$$Z(t,y) \leq C(1 + |y|^2), \ \textit{on } [0,T] \times \mathbb{R}^n,$$

then $Z \equiv Z_0$ holds in $[0,T] \times \mathbb{R}^n$.

2.4 Verification Theorem

In this subsection, we revisit the verification theorem. Define

$$\mathcal{A}_T := \left\{(c,\pi) \in \mathcal{A}_T^1; E^\pi\left[\overline{M_T^\pi}\right] = 1\right\}, \tag{18}$$

where $\overline{M_t^\pi} := \mathrm{e}^{\int_0^t (D\tilde{Z}(u,Y_u))^*\lambda(Y_u)dW_u^\pi - \frac{1}{2}\int_0^t (D\tilde{Z}(u,Y_u))^* a(Y_u) D\tilde{Z}(u,Y_u)du}$, and $\tilde{Z}(t,y)$ is the solution of (12) given in Theorem 1. We once again cite the verification theorem obtained in [1].

Theorem 2 ([1] : Theorem 4.1) *Assume* **(H1)** $\sim$ **(H3)**. *Let $\tilde{Z}$ be the unique solution of* (12) *given in Theorem* 1. *Define*

$$\widehat{c}(t,y) := \mathrm{e}^{-\frac{\tilde{Z}(t,y)}{1-\gamma}},$$

$$\widehat{\pi}(t,y) := \frac{1}{1-\gamma}(\sigma(y)\sigma(y)^*)^{-1}\left\{\mu(y) - r(y)\mathbf{1} + \sigma(y)\lambda(y)^* D\tilde{Z}(t,y)\right\}.$$

That is, $\hat{c}(t,y) = c^{\tilde{Z}}(t,y)$, $\hat{\pi}(t,y) = \pi^{\tilde{Z}}(t,y)$. Then, $(\widehat{c}_t, \widehat{\pi}_t) := (\widehat{c}(t,Y_t), \widehat{\pi}(t,Y_t)) \in \mathcal{A}_T$ is an optimal strategy for (4), *and $V(0,x,y) = \widetilde{V}(0,x,y)$. Here we define*

$$\widetilde{V}(t,x,y) := \frac{x^\gamma}{\gamma}\mathrm{e}^{\tilde{Z}(t,y)}. \tag{19}$$

3 Policy Improvement Algorithm

In this section, we present details of PIA proposed in [2]. We also recall the main theoretical results in [2], that is, the convergence of our algorithm with the true value function which is obtained in Hata et al. [1].

We define $J(t, T; c, \pi)$ which replaces J in (5) with J starting at time t from (x, y);

$$J(t, T; c, \pi) = \frac{x^\gamma}{\gamma} E_{t,y}\left[\int_t^T c_s^\gamma e^{\int_t^s [\gamma\{\ell(Y_u,\pi_u)-c_u\}-\rho]du} \tilde{M}^\pi_{t,s} ds \right. \\ \left. + e^{\int_t^T [\gamma\{\ell(Y_s,\pi_s)-c_s\}-\rho]ds} \tilde{M}^\pi_{t,T}\right], \tag{20}$$

where $\tilde{M}^\pi_{t,s} = \tilde{M}^\pi_s / \tilde{M}^\pi_t$. Define the class of the admissible strategies $\overline{\mathcal{A}}_{t,T}$ as

$$\overline{\mathcal{A}}_{t,T} := \Big\{(c_s, \pi_s)_{s\in[t,T]};\ (c_s, \pi_s) \text{ is a } [0, \infty) \\ \times\mathbb{R}^m\text{-valued } \mathcal{F}_s\text{-progressively measurable,} \\ \text{stochastic process such that } J(t, T; c, \pi) > -\infty\Big\}.$$

Take $\left(c_s^{(0)}, \pi_s^{(0)}\right) = \left(c^{(0)}(s, y), \pi^{(0)}(s, y)\right) = (1, 0)$. Using $\left(\widehat{c}_s^{(0)}, \widehat{\pi}_s^{(0)}\right) = \left(c^{(0)}(s, Y_s), \pi^{(0)}(s, Y_s)\right) = (1, 0)$, we define $V^{(0)}(t, x, y) := J(t, T; \widehat{c}^{(0)}, \widehat{\pi}^{(0)})$. Then,

$$V^{(0)}(t, x, y) = \frac{x^\gamma}{\gamma} E_{t,y}\left[\int_t^T \{\widehat{c}_s^{(0)}\}^\gamma e^{\int_0^s [\gamma\{\ell(Y_u,\widehat{\pi}_u^{(0)})-\widehat{c}_u^{(0)}\}-\rho]du} \tilde{M}^{\widehat{\pi}^{(0)}}_{t,s} ds \right. \\ \left. + e^{\int_t^T [\gamma\{\ell(Y_s,\widehat{\pi}_s^{(0)})-\widehat{c}_s^{(0)}\}-\rho]ds} \tilde{M}^{\widehat{\pi}^{(0)}}_{t,T}\right] \tag{21}$$

$$\geq \frac{x^\gamma}{\gamma}\left(1 - \frac{1}{\gamma}\right) e^{-\gamma T} > -\infty,$$

which implies $\left(\widehat{c}^{(0)}, \widehat{\pi}^{(0)}\right) \in \overline{\mathcal{A}}_{t,T}$. On the other hand, $V^{(0)}(t, x, y)$ solves

$$\frac{\partial V^{(0)}}{\partial t} + \frac{1}{2}\mathrm{tr}(a(y)D_{yy}V^{(0)}) + g(y)^* D_y V^{(0)} + \frac{\{c^{(0)}\}^\gamma x^\gamma}{\gamma} - \rho V^{(0)} \\ + x\{\pi^{(0)}\}^*\sigma(y)\lambda(y)^* D_{xy}V^{(0)} + \frac{1}{2}\{\pi^{(0)}\}^*\sigma(y)\sigma(y)^*\pi^{(0)}x^2 D_{xx}V^{(0)} \\ + [\{\pi^{(0)}\}^*(\mu(y) - r(y)\mathbf{1}) + r(y) - c^{(0)}]x D_x V^{(0)} = 0, \\ V^{(0)}(T, x, y) = \frac{x^\gamma}{\gamma}.$$

Set $V^{(0)}(t,x,y) = \dfrac{x^\gamma}{\gamma} e^{Z^{(0)}(t,y)}$. Then $Z^{(0)}(t,y)$ solves

$$\begin{aligned}
&\frac{\partial Z^{(0)}}{\partial t} + \frac{1}{2}\mathrm{tr}(a(y)D^2 Z^{(0)}) + \frac{1}{2}(DZ^{(0)})^* a(y) DZ^{(0)} + g(y)^* DZ^{(0)} - \rho \\
&- \gamma c^{(0)} + \{c^{(0)}\}^\gamma e^{-Z^{(0)}} + \gamma[\ell(y,\pi^{(0)}) + \{\pi^{(0)}\}^* \sigma(y)\lambda(y)^* DZ^{(0)}] = 0, \\
&Z^{(0)}(T,y) = 0.
\end{aligned}$$

In the sequential induction as described above, we derive each criterion $V^{(l)}$ and strategies $(c^{(l)}, \pi^{(l)})$. Define $(c^{(l)}, \pi^{(l)})$ $(l = 1, 2, \ldots)$, where

$$\begin{aligned}
c^{(l)} = c^{(l)}(t,y) &:= \mathrm{argmin}\left\{-\gamma c + c^\gamma e^{-Z^{(l-1)}}\right\} = e^{-\frac{1}{1-\gamma} Z^{(l-1)}(t,y)}, \\
\pi^{(l)} = \pi^{(l)}(t,y) &:= \mathrm{argmin}\left[\gamma\left\{\ell(y,\pi) + \pi^* \sigma(y)\lambda(y)^* DZ^{(l-1)}\right\}\right] \\
&= \frac{1}{1-\gamma}(\sigma(y)\sigma(y)^*)^{-1}\{\mu(y) - r(y)\mathbf{1} \\
&\quad + \sigma(y)\lambda(y)^* DZ^{(l-1)}(t,y)\}.
\end{aligned} \tag{22}$$

Using $\left(\widehat{c}_s^{(l)}, \widehat{\pi}_s^{(l)}\right) = \left(c^{(l)}(s,Y_s), \pi^{(l)}(s,Y_s)\right)$, we define $V^{(l)}(t,x,y) := J(t,T;\widehat{c}^{(l)}, \widehat{\pi}^{(l)})$. Then, $V^{(l)}(t,x,y)$ solves

$$\begin{aligned}
&\frac{\partial V^{(l)}}{\partial t} + \frac{1}{2}\mathrm{tr}(a(y)D_{yy}V^{(l)}) + g(y)^* D_y V^{(l)} + \frac{\{c^{(l)}\}^\gamma x^\gamma}{\gamma} - \rho V^{(l)} \\
&+ x\{\pi^{(l)}\}^* \sigma(y)\lambda(y)^* D_{xy} V^{(l)} + \frac{1}{2}\{\pi^{(l)}\}^* \sigma(y)\sigma(y)^* \pi^{(l)} x^2 D_{xx} V^{(l)} \\
&+ [\{\pi^{(l)}\}^*(\mu(y) - r(y)\mathbf{1}) + r(y) - c^{(l)}] x D_x V^{(l)} = 0,
\end{aligned}$$

with $V^{(l)}(T,x,y) = \dfrac{x^\gamma}{\gamma}$. Set $V^{(l)}(t,x,y) = \dfrac{x^\gamma}{\gamma} e^{Z^{(l)}(t,y)}$. Then $Z^{(l)}(t,y)$ solves

$$\begin{aligned}
&\frac{\partial Z^{(l)}}{\partial t} + \frac{1}{2}\mathrm{tr}(a(y)D^2 Z^{(l)}) + \frac{1}{2}(DZ^{(l)})^* a(y) DZ^{(l)} + g(y)^* DZ^{(l)} - \rho \\
&- \gamma c^{(l)} + \{c^{(l)}\}^\gamma e^{-Z^{(l)}} + \gamma[\ell(y,\pi^{(l)}) + \{\pi^{(l)}\}^* \sigma(y)\lambda(y)^* DZ^{(l)}] = 0, \\
&Z^{(l)}(T,y) = 0.
\end{aligned} \tag{23}$$

Let us proceed with the following preparations. We define the probability measure $P^{(l)}$ for each $l = 1, 2, \ldots$. as follows:

$$\left.\frac{dP^{(l)}}{dP}\right|_{\mathcal{F}_T} := M_{0,T}^{\widehat{\pi}^{(l)}}. \tag{24}$$

Here $M_{t,s}^{\widehat{\pi}^{(l)}}$ is defined by

$$\begin{aligned} M_{t,s}^{\widehat{\pi}^{(l)}} :=& \mathrm{e}^{\gamma \int_t^s \{(\widehat{\pi}_u^{(l)})^* \sigma(Y_u) - \frac{1}{1-\gamma}(DZ^{(l-1)}(u,Y_u))^* \lambda(Y_u)\} dW_u} \\ & \cdot \mathrm{e}^{-\frac{\gamma^2}{2} \int_t^s |(\widehat{\pi}_u^{(l)})^* \sigma(Y_u) - \frac{1}{1-\gamma}(DZ^{(l-1)}(u,Y_u))^* \lambda(Y_u)|^2 du}. \end{aligned} \tag{25}$$

Under $P^{(l)}$, Y_s solves

$$\begin{aligned} dY_s =& \left[B(Y_s) + Q(Y_s)DZ^{(l-1)}(s, Y_s) - \frac{1}{1-\gamma} a(Y_s)DZ^{(l-1)}(s, Y_s)\right] ds \\ & + \lambda(Y_s) dW_s^{(l)}, \quad Y_t = y, \end{aligned} \tag{26}$$

where $W_s^{(l)}$ is a Brownian motion under $P^{(l)}$ defined by

$$W_s^{(l)} = W_s - \gamma \int_t^s \left\{\sigma(Y_u)^* \widehat{\pi}_u^{(l)} - \frac{1}{1-\gamma} \lambda(Y_u)^* DZ^{(l-1)}(u, Y_u)\right\} du. \tag{27}$$

Moreover, we define $\zeta(t; s, y)$ and $\psi^{(l)}(t, y)$ by

$$\zeta(t; s, y) := E_{t,y}\left[\mathrm{e}^{\int_t^s \gamma r(Y_u) du}\right], \tag{28}$$

$$\psi^{(l)}(t, y) := E_{t,y}^{(l)}\left[\int_t^T \mathrm{e}^{\int_t^s L^{(l)}(u,Y_u) du} ds + \mathrm{e}^{\int_t^T L^{(l)}(u,Y_u) du}\right]. \tag{29}$$

Here, $E_{t,y}^{(l)}[\cdot] (l = 1, 2, \ldots)$ is the expectation with respect to the probability measure $P^{(l)}$, $L^{(1)}(t, y)$ and $L^{(l)}(t, y) (l = 2, 3, \ldots)$ is defined by

$$\begin{aligned} L^{(1)}(t, y) =& -\frac{1}{1-\gamma}(\rho + U(y)) - \frac{\gamma}{1-\gamma}\left(\gamma - \mathrm{e}^{-Z^{(0)}(t,y)}\right) - \gamma \mathrm{e}^{-\frac{1}{1-\gamma} Z^{(0)}(t,y)} \\ & + \frac{\gamma}{2(1-\gamma)^2}(DZ^{(0)}(t, y))^* \lambda(y) \\ & \left\{I - \sigma(y)^*(\sigma(y)\sigma(y)^*)^{-1}\sigma(y)\right\} \lambda(y)^* DZ^{(0)}(t, y) \\ & - \frac{\gamma^2}{2(1-\gamma)^2}\left\{\mu(y) - r(y)\mathbf{1} + \sigma(y)\lambda(y)^* DZ^{(0)}(t, y)\right\}^* \\ & \cdot (\sigma(y)\sigma(y)^*)^{-1}\left\{\mu(y) - r(y)\mathbf{1} + \sigma(y)\lambda(y)^* DZ^{(0)}(t, y)\right\} < 0, \end{aligned}$$

$$L^{(l)}(t,y) = -\frac{1}{1-\gamma}(\rho + U(y)) + \frac{\gamma}{2(1-\gamma)^2}(DZ^{(l-1)}(t,y))^*\lambda(y)$$
$$\cdot \{I - \sigma(y)^*(\sigma(y)\sigma(y)^*)^{-1}\sigma(y)\}\lambda(y)^* DZ^{(l-1)}(t,y)$$
$$- \frac{\gamma^2}{2(1-\gamma)^2}(D(Z^{(l-2)} - Z^{(l-1)})(t,y))^*\lambda(y)\sigma(y)^*$$
$$\cdot (\sigma(y)\sigma(y)^*)^{-1}\sigma(y)\lambda(y)^* D(Z^{(l-2)} - Z^{(l-1)})(t,y)$$
$$- \frac{\gamma}{1-\gamma}\left[\gamma e^{-\frac{1}{1-\gamma}Z^{(l-2)(t,y)}} - e^{-\frac{\gamma}{1-\gamma}Z^{(l-2)}(t,y) - Z^{(l-1)}(t,y)}\right]$$
$$- \gamma e^{-\frac{1}{1-\gamma}Z^{(l-1)}(t,y)} < 0.$$

Here, we introduce the results of our algorithm obtained in [2].

Theorem 3 ([2]: Theorem 4) *Assume* **(H1)** $\sim$ **(H3)**. *Assume also one of* **(H4a)** *or* **(H4b)**.

(H4a) $\gamma \in (-\infty, 0)$ *and* $U(y)$ *is bounded,*
(H4b) $\gamma \in (-1, 0)$.

Define

$$\left(\widehat{c}_s^{(l)}, \widehat{\pi}_s^{(l)}\right) := \begin{cases} (1,0) & (l=0), \\ \left(c^{(l)}(s,Y_s), \pi^{(l)}(s,Y_s)\right) & (l=1,2,3,\ldots.), \end{cases} \tag{30}$$

where $c^{(l)}(t,y)$ *and* $\pi^{(l)}(t,y)$ *defned by* (22). *Then, for each* $l = 0,1,2,\ldots,$ $\left(\widehat{c}_s^{(l)}, \widehat{\pi}_s^{(l)}\right) \in \overline{\mathcal{A}}_{t,T}$ *and* $Z^{(l)} \in C^{1,2}((0,T)\times\mathbb{R}^n)\cap C([0,T]\times\mathbb{R}^n)$ *holds. Moreover, we have*

$$Z^{(l)}(t,y) = -\gamma \log x + \log\{\gamma J(t,T;\widehat{c}^{(l)}, \widehat{\pi}^{(l)})\}$$
$$= \begin{cases} \log\left(\int_t^T e^{-(\gamma+\rho)(s-t)}\zeta(t;s,y)ds + e^{-(\gamma+\rho)(T-t)}\zeta(t;T,y)\right), & (l=0), \\ -\frac{\gamma}{1-\gamma}Z^{(l-1)}(t,y) + \log\psi^{(l)}(t,y), & (l=1,2,\ldots.). \end{cases} \tag{31}$$

Next, we introduce a lemma to prove the convergence of our algorithm.

Lemma 1 ([2]: Lemma 2) *Assume* **(H1)** $\sim$ **(H3)**. *Assume also one of* **(H4a)** *or* **(H4b)**.

1. *For each* $l = 0,1,2,\ldots,$ $Z^{(l)}(t,y) \geq \tilde{Z}(t,y)$ *holds on* $[0,T]\times\mathbb{R}^n$.
2. *For each* $l = 1,2,\ldots,$ $(1-\gamma)Z^{(l)}(t,y) + \gamma Z^{(l-1)}(t,y) \leq \tilde{Z}(t,y)$ *holds on* $[0,T]\times\mathbb{R}^n$.

Using Lemma 1 and noting $0 < \dfrac{-\gamma}{1-\gamma} < 1$, we have

$$0 \le Z^{(l)} - \tilde{Z} \le \left(\frac{-\gamma}{1-\gamma}\right)\left(Z^{(l-1)} - \tilde{Z}\right) \le Z^{(l-1)} - \tilde{Z}.$$

So, we see that for $l = 1, 2, \ldots$, $Z^{(l)} \le Z^{(l-1)}$ holds, which means that $Z^{(l)}$ is monotonic with respect to l. Further, repeating this argument, [2] obtained the following theorem of the convergence of our algorithm.

Theorem 4 ([2]: Theorem 5) *Assume* **(H1)** $\sim$ **(H3)**. *Assume also one of* **(H4a)** *or* **(H4b)**. *Then, the following assertions hold.*

1. *The following relations hold :*

$$Z^{(0)}(t, y) \ge Z^{(1)}(t, y) \ge \ldots \ge Z^{(l)}(t, y) \ge \ldots \ge \tilde{Z}(t, y),$$

$$0 \le Z^{(l)}(t, y) - \tilde{Z}(t, y) \le \left(\frac{-\gamma}{1-\gamma}\right)^{l}\left(Z^{(0)}(t, y) - \tilde{Z}(t, y)\right) \text{ on } [0, T] \times \mathbb{R}^n,$$

which implies $Z^{(l)}(t, y) \downarrow \tilde{Z}(t, y)$ as $l \to \infty$.

2. *The following relations hold :*

$$V^{(0)}(t, x, y) \le V^{(1)}(t, x, y) \le \ldots \le V^{(l)}(t, x, y) \le \ldots \le \tilde{V}(t, x, y),$$

or equivalently

$$\begin{aligned} J(t, T; \widehat{c}^{(0)}, \widehat{\pi}^{(0)}) &\le J(t, T; \widehat{c}^{(1)}, \widehat{\pi}^{(1)}) \le \ldots \\ &\le J(t, T; \widehat{c}^{(l)}, \widehat{\pi}^{(l)}) \le \ldots \le J(t, T; \widehat{c}, \widehat{\pi}) = \tilde{V}(t, x, y). \end{aligned}$$

Futher, the strategy $\left(\widehat{c}_t^{(l)}, \widehat{\pi}_t^{(l)}\right)$ *is nearly optimal when* l *is sufficiently large. More precisely, we have the following.*

$$\begin{aligned} \lim_{l\to\infty} J\left(t, T; \widehat{c}^{(l)}, \widehat{\pi}^{(l)}\right) &= \lim_{l\to\infty} V^{(l)}(t, x, y) = \lim_{l\to\infty} \frac{x^\gamma}{\gamma} \mathrm{e}^{Z^{(l)}(t,y)} = \frac{x^\gamma}{\gamma} \mathrm{e}^{\tilde{Z}(t,y)} \\ &= \tilde{V}(t, x, y), \end{aligned}$$

where $\tilde{V}(t, x, y)$ *is defined by* (19). *So, we can regard* $\left\{\left(\widehat{c}^{(l)}, \widehat{\pi}^{(l)}\right)\right\}_{l\in\{0\}\cup\mathbb{N}}$ *as PIA.*

4 Numerical Results

In this section, we will provide some numerical results of our PIA to understand the convergence behavior. We assume the following.

(N1) $m = n = 1$ (one risky asset and one factor process),
(N2) $r(y) = r,\ \mu(y) := a_0 + a_1 \tan^{-1} y,\ g(y) := b_0 + b_1 \tan^{-1} y,\ \lambda = \Lambda, \sigma := \Sigma$, where the followings are independent of y:

$$r \geq 0,\ a_0, a_1, b_0, b_1 \in \mathbb{R},\ \Lambda,\ \Sigma \in \mathbb{R}^2.$$

We will utilize the parameter values for the sequential numerical simulation.

Furthermore, the time until maturity is discretized into $n_{time} = 1000$ equal intervals, and set $\Delta t = T/n_{time}$. And similarly, the factor state space is also discretized from -5 to 5 into $n_{sp} = 2000$ equal intervals, and set $\Delta y = (5 - (-5))/n_{sp}$. In the following numerical experiments, we use the finite difference method with those discretization to simulate PDEs which appear in PIA.

In this setting, we note that

$$U(y) = -\frac{\gamma}{2(1-\gamma)}(\Sigma\Sigma^*)^{-1}(a_0 - r + a_1 \tan^{-1} y)^2 - \gamma r$$

is bounded. So, this setting satisfies **(H4a)**. Furthermore, we note that $Q(y)$ and $B(y)$ turn to be the following:

$$Q(y) = Q := \Lambda\left(I + \frac{\gamma}{1-\gamma}\Sigma^*(\Sigma\Sigma^*)^{-1}\Sigma\right)\Lambda^*,$$

$$B(y) := (b_0 + b_1 \tan^{-1} y) + \frac{\gamma}{1-\gamma}\Lambda\Sigma^*(\Sigma\Sigma^*)^{-1}(a_0 - r + a_1 \tan^{-1} y).$$

Applying **(N1)** and **(N2)** to the results obtained in Theorem 4, we recall $Z^{(l)} (l = 0, 1, 2, \ldots,)$ is represented as follows:

$$Z^{(0)}(t, y) = \log\left(\int_t^T e^{-(\gamma+\rho+\gamma r)(s-t)} ds + e^{-(\gamma+\rho+\gamma r)(T-t)}\right),$$

$$Z^{(l)}(t, y) = -\frac{\gamma}{1-\gamma} Z^{(l-1)}(t, y) + \log \psi^{(l)}(t, y),\ \ l = 1, 2, \ldots,$$

where $\psi^{(l)}(t, y)$, $L^{(1)}(t, y)$ and $L^{(l)}(t, y)$ are defined as follows:

$$\psi^{(l)}(t, y) := E_{t,y}^{(l)}\left[\int_t^T e^{\int_t^s L^{(l)}(u, Y_u)du} ds + e^{\int_t^T L^{(l)}(u, Y_u)du}\right], \tag{32}$$

$$
\begin{aligned}
L^{(1)}(t,y) :=& -\frac{1}{1-\gamma}(\rho + U(y)) - \frac{\gamma}{1-\gamma}\left(\gamma - \mathrm{e}^{-Z^{(0)}(t,y)}\right) \\
& - \gamma \mathrm{e}^{-\frac{1}{1-\gamma}Z^{(0)}(t,y)} < 0, \\
L^{(l)}(t,y) :=& -\frac{1}{1-\gamma}(\rho + U(y)) \\
& + \frac{\gamma}{2(1-\gamma)^2}\Lambda\{I - \Sigma^*(\Sigma\Sigma^*)^{-1}\Sigma\}\Lambda^*(DZ^{(l-1)}(t,y))^2 \\
& - \frac{\gamma^2}{2(1-\gamma)^2}\Lambda\Sigma^*(\Sigma\Sigma^*)^{-1}\Sigma\Lambda^*(D(Z^{(l-2)} - Z^{(l-1)})(t,y))^2 \\
& - \frac{\gamma}{1-\gamma}\left[\gamma \mathrm{e}^{-\frac{1}{1-\gamma}Z^{(l-2)}(t,y)} - \mathrm{e}^{-\frac{\gamma}{1-\gamma}Z^{(l-2)}(t,y) - Z^{(l-1)}(t,y)}\right] \\
& - \gamma \mathrm{e}^{-\frac{1}{1-\gamma}Z^{(l-1)}(t,y)} < 0, \quad \text{for } l = 2, 3, \ldots.
\end{aligned}
$$

Here $E^{(l)}_{t,y}[\cdot]$ is the expectation with respect to the probability measure $P^{(l)}$ defined by (24) Recall that under $P^{(l)}$, Y_s solves

$$
dY_s = \left[B(Y_s) + \left\{Q - \frac{1}{1-\gamma}\Lambda\Lambda^*\right\}DZ^{(l-1)}(s, Y_s)\right]ds + \Lambda dW^{(l)}_s,\; Y_t = y,
$$

where $W^{(l)}_s$ is a Brownian motion under $P^{(l)}$ and is given in (27). We observe that $\psi^{(l)}$ is a unique classical solution of the linear partial differential equation:

$$
\begin{aligned}
& \frac{\partial \psi}{\partial t} + \frac{1}{2}\mathrm{tr}(\Lambda\Lambda^* D^2\psi) + (D\psi)^*\left[B(y) + \left\{Q - \frac{1}{1-\gamma}\Lambda\Lambda^*\right\}\right. \\
& \left.\cdot DZ^{(l-1)}(t,y)\right] + L^{(l)}(t,y)\psi = -1, \text{ on } (0,T)\times\mathbb{R}^n, \\
& \psi(T,y) = 1.
\end{aligned} \tag{33}
$$

In Theorem 1 of [3], (34) is approximated by a sequence of classical solutions $w_R(t,y)$ for linear partial differential equations :

$$
\begin{aligned}
& \frac{\partial w}{\partial t} + \frac{1}{2}\mathrm{tr}(\Lambda\Lambda^* D^2 w) + (Dw)^*\left[B(y) + \left\{Q - \frac{1}{1-\gamma}\Lambda\Lambda^*\right\}DZ^{(l-1)}(t,y)\right] \\
& + L^{(l)}(t,y)w = -1, \quad \text{on } (0,T)\times B_R, \\
& w(t,y) = \psi^{(l)}(t,y), \quad \text{on } (0,T)\times\partial B_R \cup \{T\}\times\overline{B}_R,
\end{aligned} \tag{34}
$$

where $B_R = \{y \in \mathbb{R}^n;\, |y| < R\}$.

4.1 About Boundary Conditions

In this subsection, we will present simulation results to establish boundary values for the simulation of partial differential equations (34) in PIA. Our PDEs are defined on the whole space in y. However, obviously we cannot simulate PDEs on the whole space on computer. So, we need to truncate the space at a certain region and set some conditions on the boundaries. Here, we adopt the idea which has been introduced in [2] and give precise illustrations which have not been mentioned in [2] due to the page restriction. The idea is that we use the Feynman-Kac representation (32) of $\psi^{(l)}(t, y)$ to obtain the boundary values of $Z^{(l)}(t, y)$, that is, we simulate $\psi^{(l)}(t, y_d)$ and $\psi^{(l)}(t, y_u)$ using the Monte-Carlo method with 10^3 iterations, where y_u (y_d) is the upper (lower) boundary value of the factor process $Y = \{Y_t\}$. Because in our proof of the convergence given in [2], we use the results obtained by [3], and it is theoretically guarantees that by taking the boundary condition like (32), its solution converges to the solution of the original PDE.

In order to carry out the above idea, first we need to truncate the factor state space. In Table 1, we assign 5 and -5 as y_u and y_d respectively. Figure 1 depicts the 10^3 sample paths of the factor process with their initial factor value $y = 0$. Figure 2 depicts the 10^3 sample paths of the factor process with their initial factor value $y = y_u$. Note that we can easily infer the paths in the case $y = y_d$. And, Fig. 3 depicts the 10^3 sample paths of the modified factor process with their initial factor value $y = y_u$, where the factor value is replaced by $y_u = 5$ when it exceeds 5. From our setting $g(y) = b_0 + b_1 \tan^{-1} y = -\tan^{-1} y$ of the drift term of the factor process and the choice of coefficient values in Table 1, the process $Y = \{Y_t\}$ has

Table 1 Parameter values

Parameter	Value	Parameter	Value
S_0^0	1	y	0
r	0.01	b_0	0
S_0	100	b_1	-1.0
a_0	0.1	Λ_1	0
a_1	0.5	Λ_2	0.1
Σ_1	$0.2 \times \sqrt{1-(-0.5)^2}$	x	1
Σ_2	$0.2 \times (-0.5)$	T	0.1
γ	-0.5	y_d	-5
ρ	0.1	y_u	5

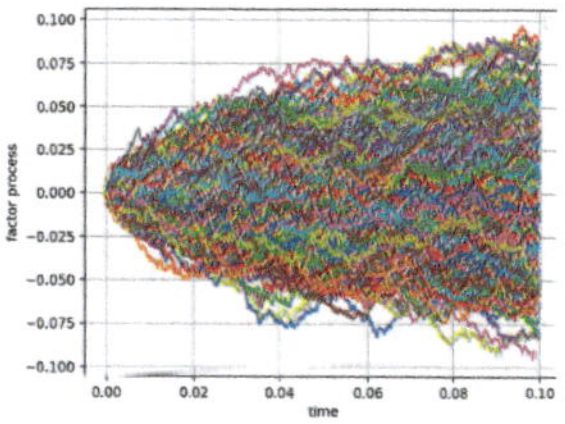

Fig. 1 Paths of factor process with $Y_0 = y$

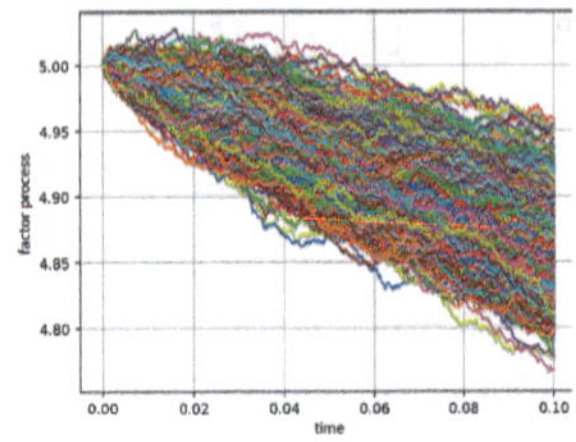

Fig. 2 Paths of factor process with $Y_0 = y_u$

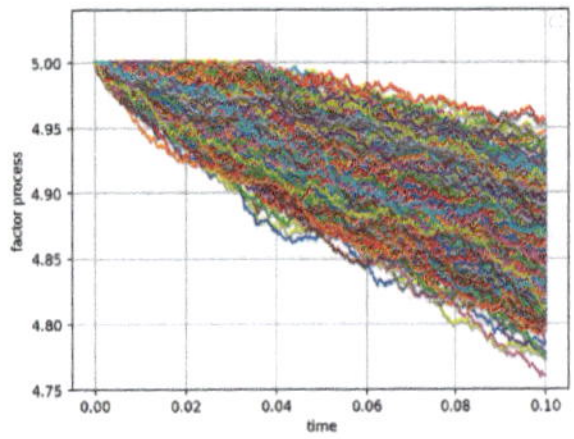

Fig. 3 Modified paths of factor process with $Y_0 = y_u$

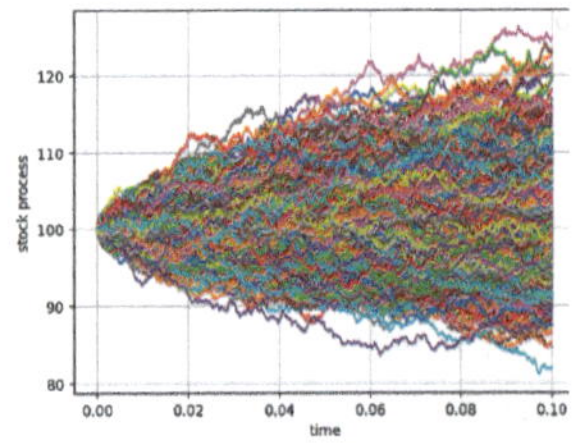

Fig. 4 Paths of stock process with $Y_0 = y_0$

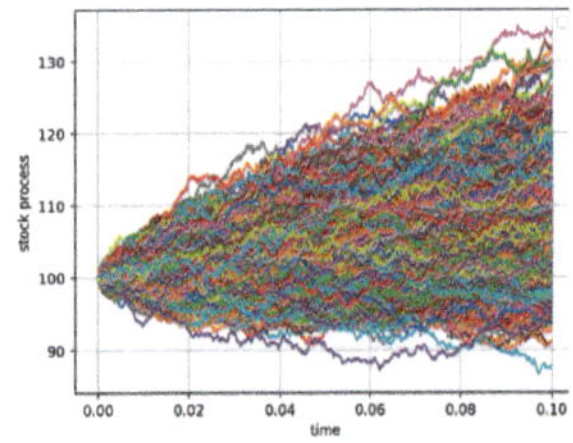

Fig. 5 Paths of stock process with $Y_0 = y_u$

a kind of mean-reverting property to level 0. When the initial value of Y_0 is much different from the mean-reverting level 0, the reverting power is quite strong and it rapidly returns to the level. Therefore, the probability that paths become larger than 5 for the upper boundary (less than -5 for the lower boundary) is regarded as close to zero. In fact, Figs. 4, 5, and 6 provide the corresponding 10^3 sample paths of the stock price process for three cases. As we can observe from Figs. 5 and 6, their sample paths and the distribution are similar even if we truncate or not at the upper boundary. Hence, we use the modified paths to calculate the upper and lower boundary values of $Z^{(l)}(t, y)$ for $l = 1, 2, 3, \ldots$ in the later simulation of $\psi^{(l)}$ using (32).

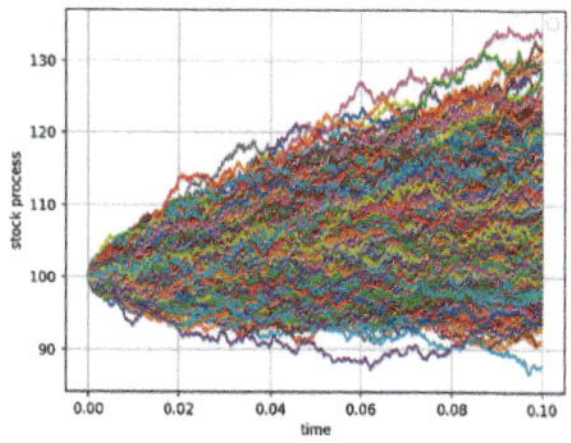

Fig. 6 Modified paths of stock process with $Y_0 = y_u$

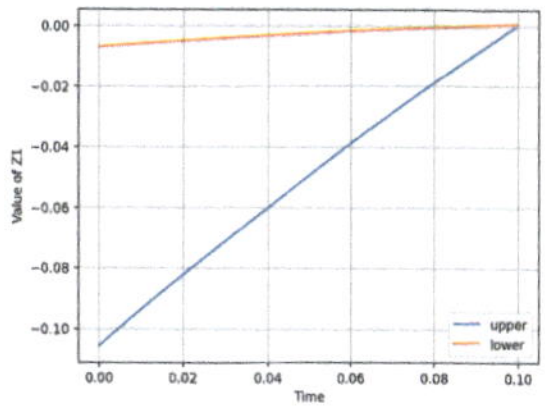

Fig. 7 Upper and lower boundary conditions: $Z^{(1)}(t, y_u)$, $Z^{(1)}(t, y_d)$

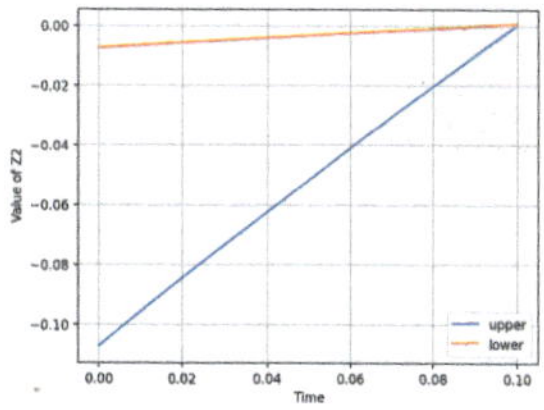

Fig. 8 Upper and lower boundary conditions: $Z^{(2)}(t, y_u)$, $Z^{(2)}(t, y_d)$

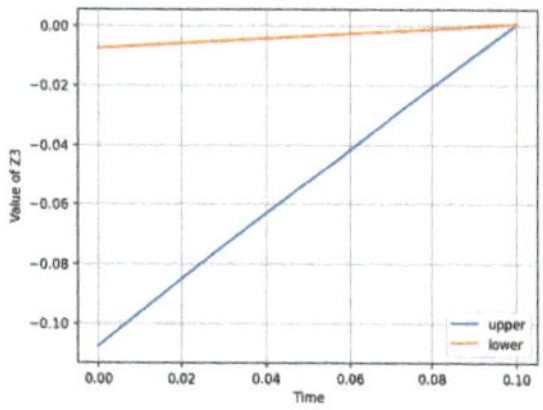

Fig. 9 Upper and lower boundary conditions: $Z^{(3)}(t, y_u)$, $Z^{(3)}(t, y_d)$

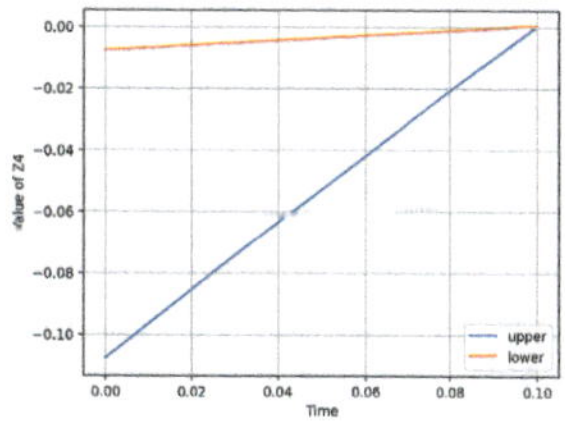

Fig. 10 Upper and lower boundary conditions: $Z^{(4)}(t, y_u)$, $Z^{(4)}(t, y_d)$

Next, we consider the upper and lower boundary values for each $Z^{(l)}(t, y)$ $(l = 1, 2, \ldots)$ in PIA. Note that for $Z^{(0)}(t, y)$, we do not need to discuss its boundary conditions because we can have its explicit expression of the solution and calculate values on the whole space. From Figs. 7, 8, 9, 10, 11, and 12, we present upper and

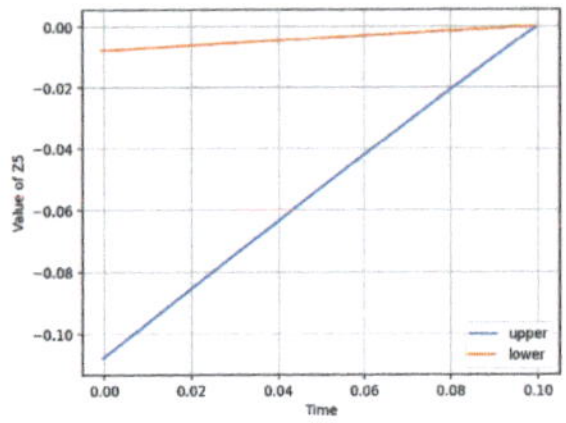

Fig. 11 Upper and lower boundary conditions: $Z^{(5)}(t, y_u)$, $Z^{(5)}(t, y_d)$

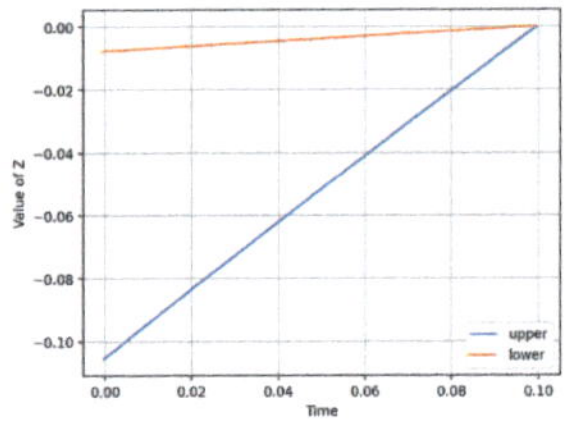

Fig. 12 Upper and lower boundary conditions: $\tilde{Z}(t, y_u)$, $\tilde{Z}(t, y_d)$

lower boundary values obtained through our following algorithm. For calculation of boundary values of $Z^{(1)}(t, y)$, we use the Feynman-Kac representation (32) with $Z^{(0)}(t, y)$. Precisely speaking, the Feynman-Kac representation is rewritten under the original probability P. Then, as mentioned above, we can simulate $Z^{(0)}(t, y)$ on the whole space due to its explicit expression and independence of y. Thanks to the expression, here, we do not need to introduce the truncation which has been mentioned above. So, for all time $t = i\Delta t$ $(i = 0, 1, \ldots, n_{time})$, we generate 10^3 sample paths of the factor process (1) starting from $Y_0 = y_u$ under our original probability P using the Euler-Maruyama approximation and obtain an approximated value of the expectation using the Monte-Carlo method. Precisely, we simulate the Feynman-Kac formula for all t at y_u. However, we can observe that the boundary values of $\psi^{(1)}(t, y)$ are relatively linear in t, so to have smooth boundaries we adopt a linear line which passes two points $(t, \psi^{(1)}) = (0, \bar{\psi}^{(1)}(0, y_u))$ and $(0.1, 1)$, where $\bar{\psi}^{(1)}(0, y_u)$ denotes the obtained Monte-Carlo values with $y = y_u$ at time $t = 0$. Such boundary lines are provided in Fig. 7. Similarly, we obtain lower boundary conditions for $Z^{(1)}$. On the other hand, to calculate boundary values of $Z^{(l)}(t, y)$ $(l = 2, 3, \ldots.)$, the truncation is needed for the factor processes when it exceeds 5 or falls below -5. Others are similar to the case of calculation of $\psi^{(1)}(t, y)$. Note that $\psi^{(l)}(t, y)$ is simulated using $DZ^{(l-2)}(t, y)$ and $DZ^{(l-1)}(t, y)$, however $Z^{(l-2)}(t, y)$ and $Z^{(l-1)}(t, y)$ are obtained only on $(t, y) = (i\Delta t, y_d + j\Delta y)$ $(i = 0, 1, \ldots, n_{time}$ and $j = 0, 1, \ldots, n_{sp})$, so when generated factor values $Y_{i\Delta t}$ do not pass on such points, then we replace it with the nearest lower value on $(t, y) = (i\Delta t, y_d + j^*\Delta y)$ where $y_d + j^*\Delta y \le Y_{i\Delta t} < y_d + (j^* + 1)\Delta y$ holds. Each boundary shape is shown in Figs. 8, 9, 10, and 11 for $l = 2, 3, 4, 5$. Boundary conditions for $\tilde{Z}(t, y)$ are simulated using the subsolution $\hat{Z}$ (14) of $\tilde{Z}$ which is obtained in [1]. In this case we generate 10^3 paths of the process $\bar{Y}$ under the measure P. Its numerical results are given in Fig. 12.

From Figs. 7, 8, 9, 10, and 11, the upper and lower boundary values are improved in the iteration. We can observe that the boundary values of $Z^{(4)}$ and $Z^{(5)}$ are almost the same. On the other hand, the boundary values of $\tilde{Z}$ are obtained using the expression of its subsolution (14), so the figures of boundary values of $Z^{(5)}$ are similar to those of $\tilde{Z}$, but they are not the same.

4.2 About Value Functions $Z^{(l)}(t, y)$ and $V^{(l)}(t, y)$ in PIA

Based on the boundary conditions obtained in the previous subsection, we simulate $Z^{(l)}$ in PIA using the finite difference scheme in this subsection, and compare them to $\tilde{Z}$ obtained by the finite difference scheme for PDE in [1]. Their results are shown in Figs. 13, 14, 15, 16, 17, 18, and 19. Here, note that the region of y in each graph is restricted from -3.75 to 3.75. Because we adopt the finite difference scheme to obtain the numerical solution of $Z^{(l)}$ and $\tilde{Z}$, then as a common problem in the finite difference scheme, some spikes appear at the boundary. But, through our numerical experiments, the impacts of those spikes on our numerical results are almost nothing.

Figures 13, 14, 15, 16, 17, and 18 show the graphs of $Z^{(l)}(t, y)$ $(l = 0, 1, \dots, 5)$, and Fig. 19 is the graph of $\tilde{Z}(t, y)$. The graph of $Z^{(0)}$ is much different from that of $\tilde{Z}$, and the graph of $Z^{(1)}$ is slightly different from that of $\tilde{Z}$. However,

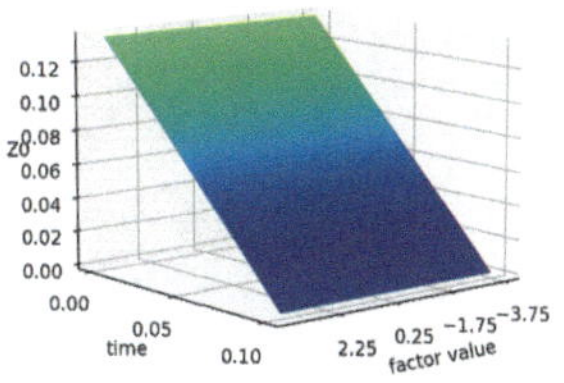

Fig. 13 Shape of $Z^{(0)}(t, y)$

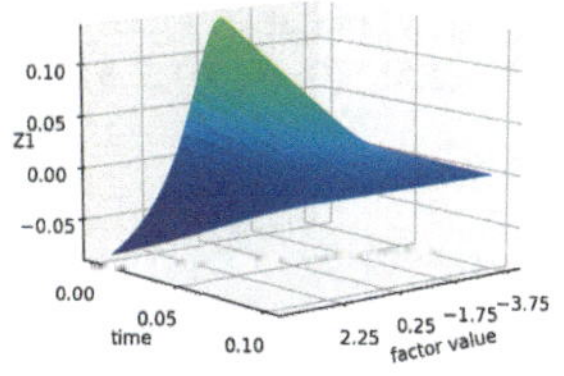

Fig. 14 Shape of $Z^{(1)}(t, y)$

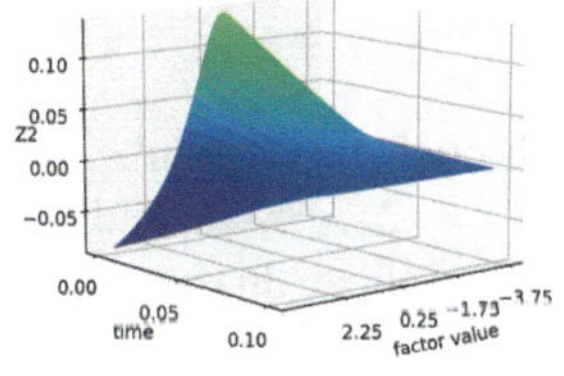

Fig. 15 Shape of $Z^{(2)}(t, y)$

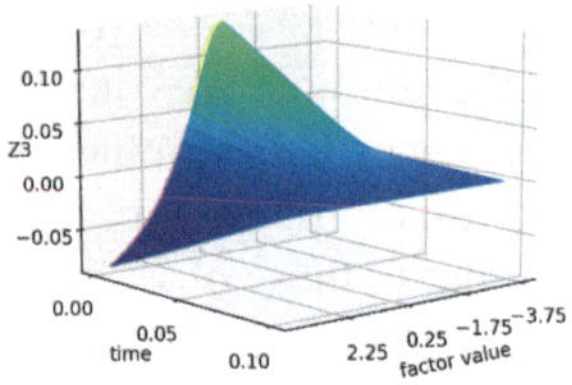

Fig. 16 Shape of $Z^{(3)}(t, y)$

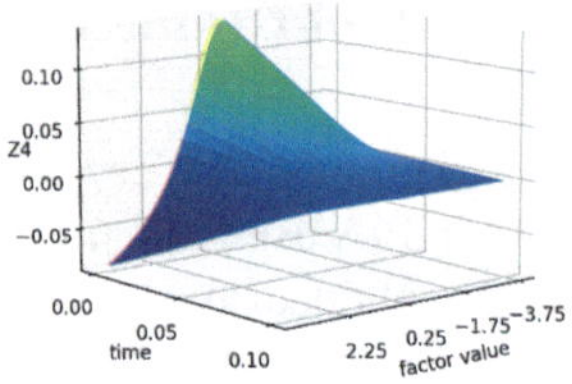

Fig. 17 Shape of $Z^{(4)}(t, y)$

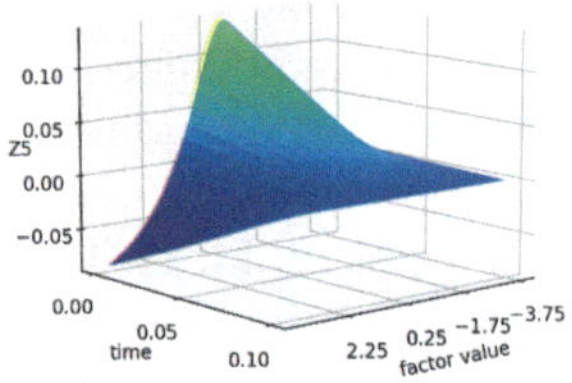

Fig. 18 Shape of $Z^{(5)}(t, y)$

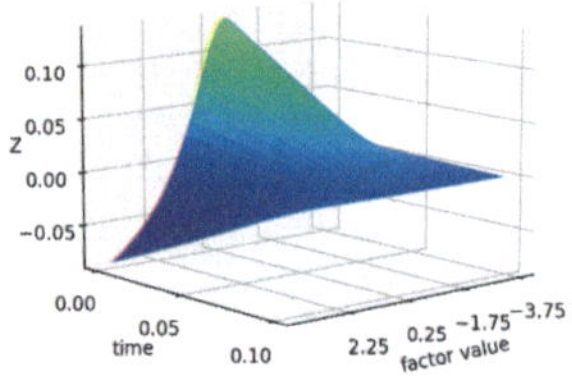

Fig. 19 Shape of $\tilde{Z}(t, y)$

Table 2 Convergence of $Z^{(l)}(t, y)$ in l at $(t, y) = (0, -2.5), (0, 0), (0, 2.5)$

l	$Z^{(l)}(0, -2.5)$	$Z^{(l)}(0, 0)$	$Z^{(l)}(0, 2.5)$
0	0.133992222	0.133992222	0.133992222
1	0.033545374	0.129982970	−0.050723785
2	0.033280779	0.129770876	−0.051121358
3	0.033210602	0.129700211	−0.051190583
4	0.033187209	0.129676656	−0.051213658
5	0.033179412	0.129668804	−0.051221350
$\tilde{Z}$	0.033175898	0.129670218	−0.051224436

through those graphs, we can not find differences among the graphs of $Z^{(l)}(t, y)$ $(l = 2, 3, 4, 5)$, and also they are quite similar to the graph of $\tilde{Z}$ on graphs. To analyze this point in more detail, Table 2 shows values of $Z^{(l)}$ and $\tilde{Z}$ at 3 points $(t, y) = (0, -2.5), (0, 0), (0, 2.5)$. In fact, those points $Z^{(l)}(t, y)$ have

Fig. 20 Shape of $V^{(0)}(t, y)$

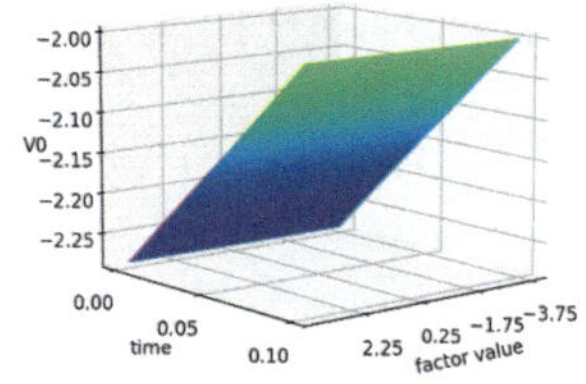

Fig. 21 Shape of $V^{(1)}(t, y)$

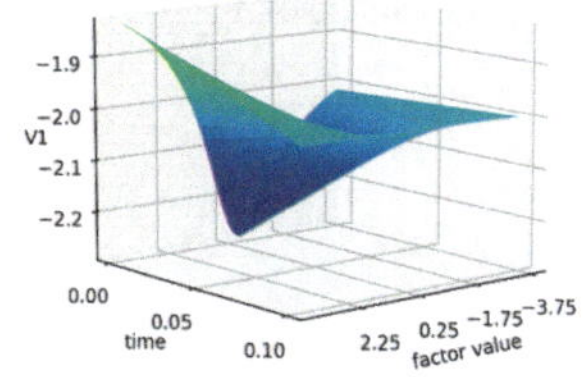

Fig. 22 Shape of $V^{(2)}(t, y)$

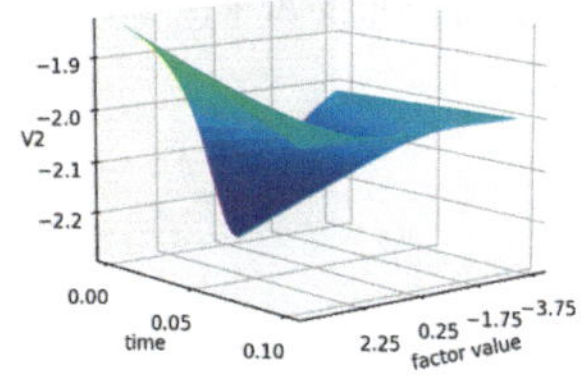

Fig. 23 Shape of $V^{(3)}(t, y)$

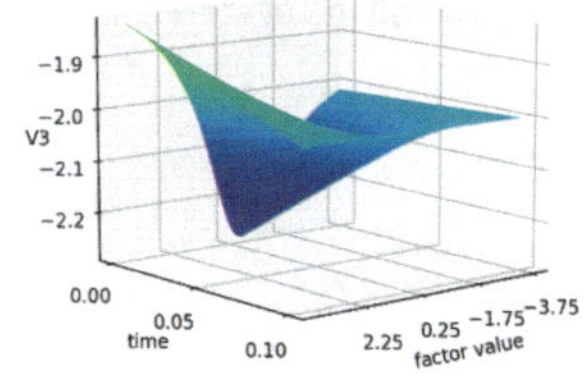

not converged to $\tilde{Z}(t, y)$ exactly in only 5 iterations, but their values $Z^{(l)}(t, y)$ are relatively close to values of $\tilde{Z}(t, y)$ even if the number of iteration is small. Note that the monotone decreasing of $Z^{(l)}(t, y)$ in l for each (t, y), which is mentioned in Theorem 4.**1**, holds here.

Next, we discuss the convergence behavior of $V^{(l)}(t, x, y)$. Actually, $V^{(l)}(t, x, y)$ is calculated through $V^{(l)}(t, x, y) = \dfrac{x^{\gamma}}{\gamma} e^{Z^{(l)}(t,y)}$, so the behavior of the convergence is very similar to that for $Z^{(l)}(t, y)$. From Table 1, now our x is 1. Figures 20, 21, 22, 23, 24, 25, and 26 are numerical results for $V^{(l)}(t, 1, y)$ and $\tilde{V}(t, 1, y)$. In fact, as before, $V^{(0)}$ and $V^{(1)}$ are some differences from $\tilde{V}$. On the other hand, for $l = 2, 3, 4, 5$, $V^{(l)}$ is very close to $\tilde{V}$ on each point (t, y).

Fig. 24 Shape of $V^{(4)}(t, y)$

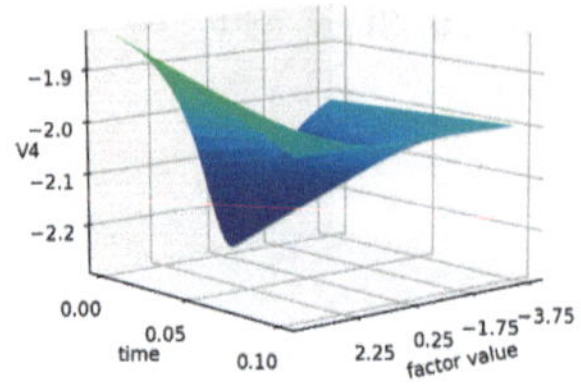

Fig. 25 Shape of $V^{(5)}(t, y)$

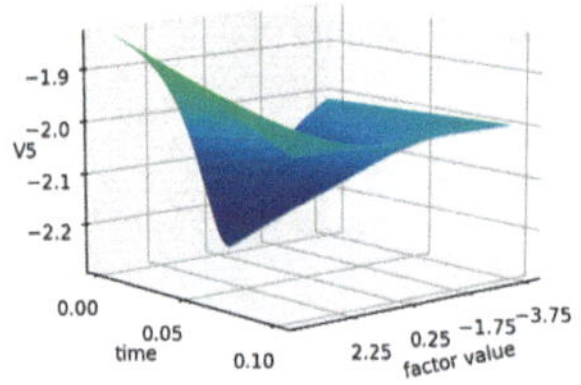

Fig. 26 Shape of $\tilde{V}(t, y)$

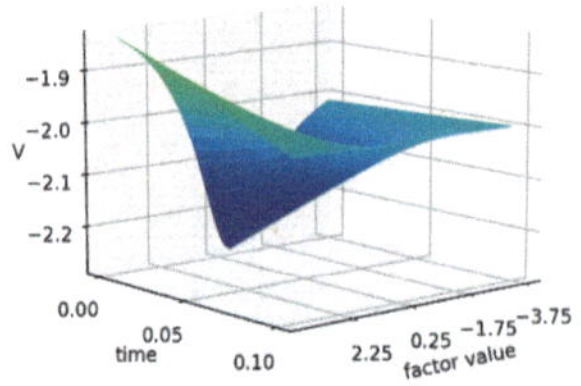

Fig. 27 Shape of $c^{(0)}(t, y)$

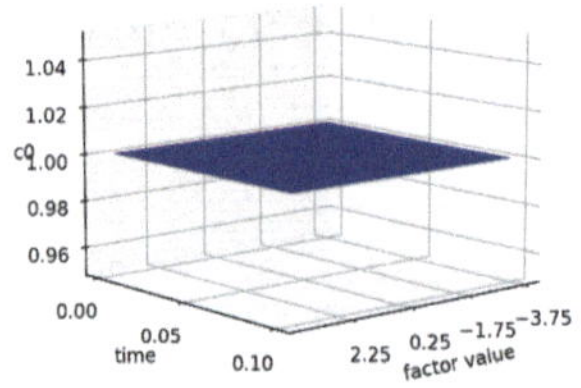

Fig. 28 Shape of $c^{(1)}(t, y)$

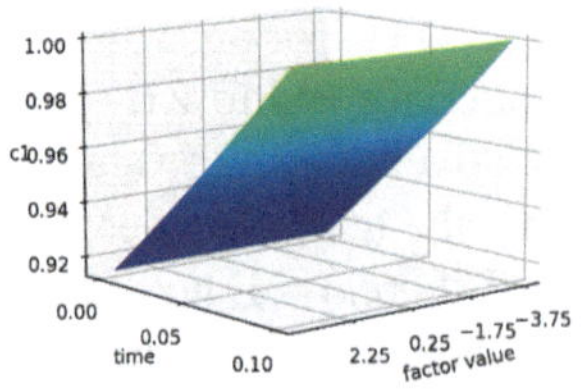

4.3 About Strategies $c^{(l)}(t, y)$ and $\pi^{(l)}(t, y)$ in PIA

In this subsection, we consider our strategies $c^{(l)}(t, y)$ and $\pi^{(l)}(t, y)$ in iteration obtained by (22).

First, in Figs. 27, 28, 29, 30, 31, and 32, we present graphs of our consumption strategy in each step, and in Fig. 33, we also provide the graph of the optimal

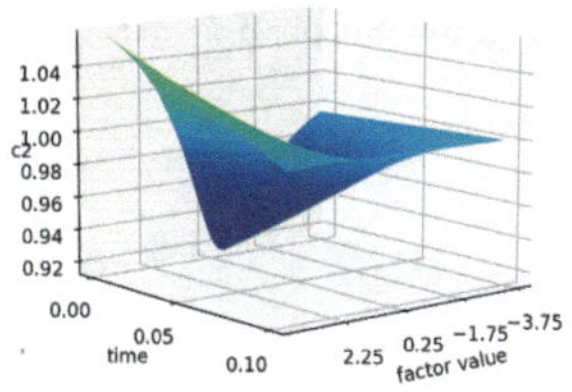

Fig. 29 Shape of $c^{(2)}(t, y)$

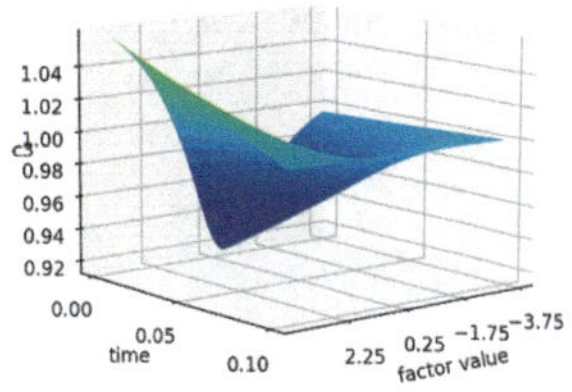

Fig. 30 Shape of $c^{(3)}(t, y)$

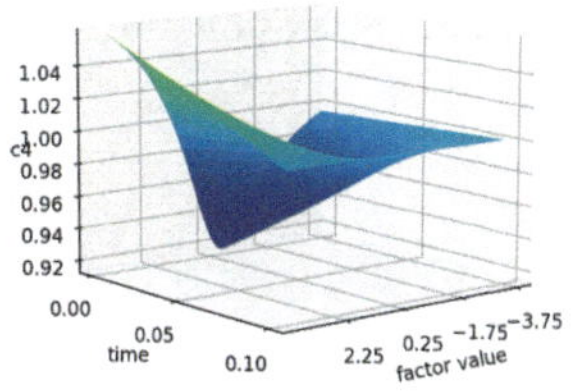

Fig. 31 Shape of $c^{(4)}(t, y)$

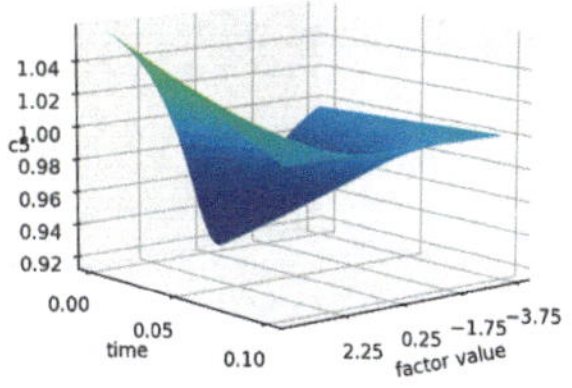

Fig. 32 Shape of $c^{(5)}(t, y)$

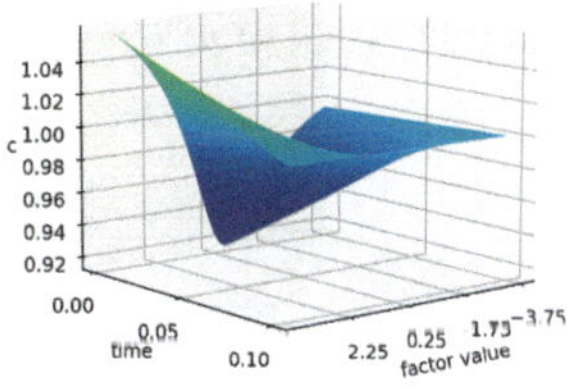

Fig. 33 Shape of $\widehat{c}(t, y)$

consumption strategy $\widehat{c}$ calculated by Theorem 2. As we can see from (22), $c^{(l)}$ is calculated using $Z^{(l-1)}$, so the convergence in l is one step late comparing to convergence of $Z^{(l)}$ in l. Hence, $c^{(l)}$ $(l = 0, 1, 2)$ have some differences from $\widehat{c}$, and $c^{(l)}$ $(l = 3, 4, 5)$ are similar to each other and $\widehat{c}$.

Next, we consider the graphs of our investment strategy in iteration obtained using (22). Figures 34, 35, 36, 37, 38, and 39 show shapes of $\pi^{(l)}$ for $l = 0, 1, \ldots, 5$, and Fig. 40 shows the shape of $\widehat{\pi}$. Obviously, $\pi^{(0)}$ is much different from others,

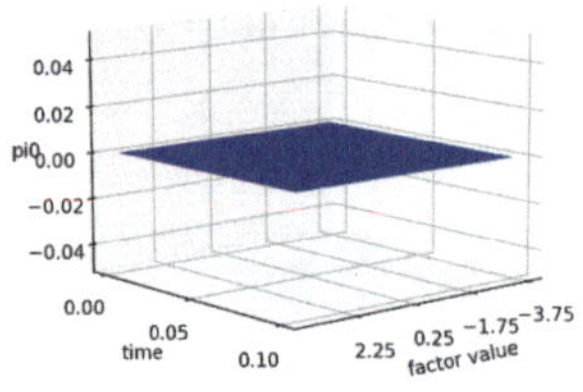

Fig. 34 Shape of $\pi^{(0)}(t, y)$

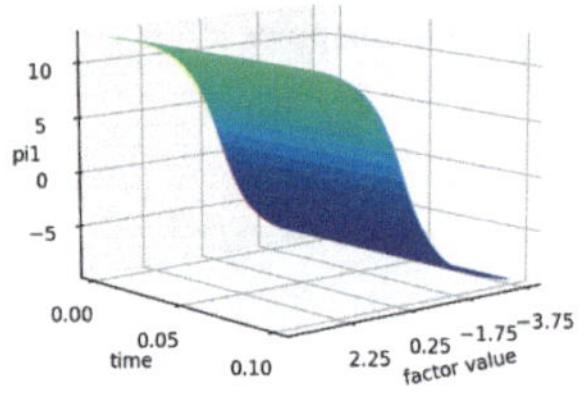

Fig. 35 Shape of $\pi^{(1)}(t, y)$

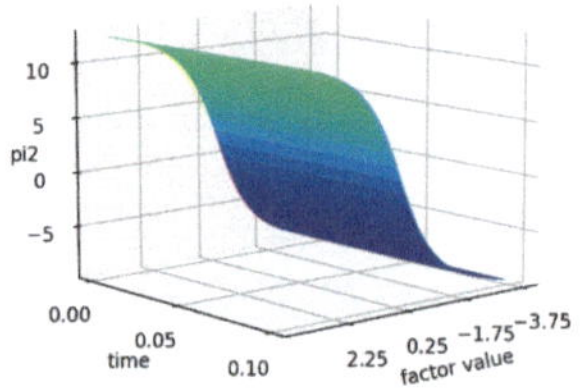

Fig. 36 Shape of $\pi^{(2)}(t, y)$

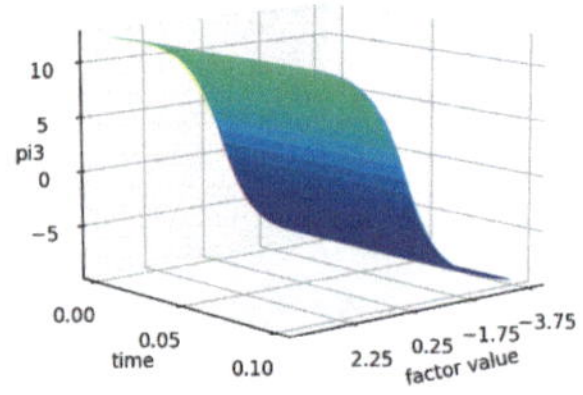

Fig. 37 Shape of $\pi^{(3)}(t, y)$

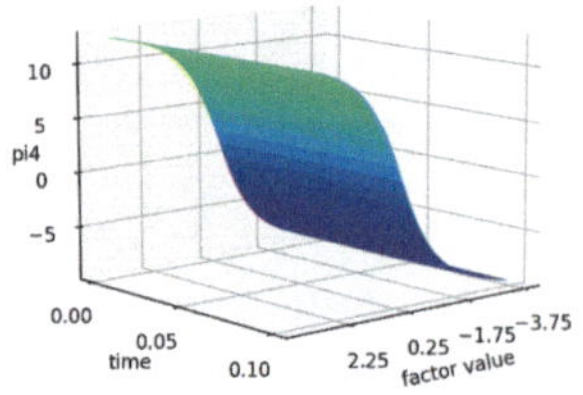

Fig. 38 Shape of $\pi^{(4)}(t, y)$

however here $\pi^{(1)}$ is already close to $\widehat{\pi}$. $\pi^{(l)}$ is also calculated using $Z^{(l-1)}$, so as $c^{(l)}$, the convergence in l should be one step late compared to $Z^{(l)}$, however it does not happen. Because from Table 1, now $\sigma\lambda^* = 0.2 \times (-0.5) \times 0.1 = -0.01$ and $\sigma\lambda^*$ is the coefficient of $DZ^{(l-1)}$, so in our parameter values here, the value of $\sigma\lambda^* DZ^{(l-1)}(t, y)$ is much smaller than the value of $\mu(y) - r$. Therefore, $\pi^{(l)}$ converges to $\tilde{\pi}$ faster than other calculations.

Fig. 39 Shape of $\pi^{(5)}(t, y)$

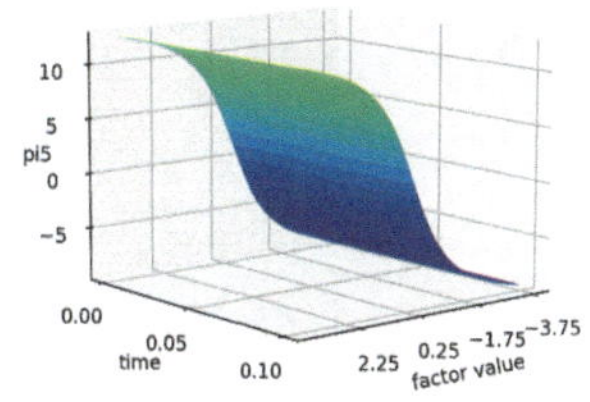

Fig. 40 Shape of $\widehat{\pi}(t, y)$

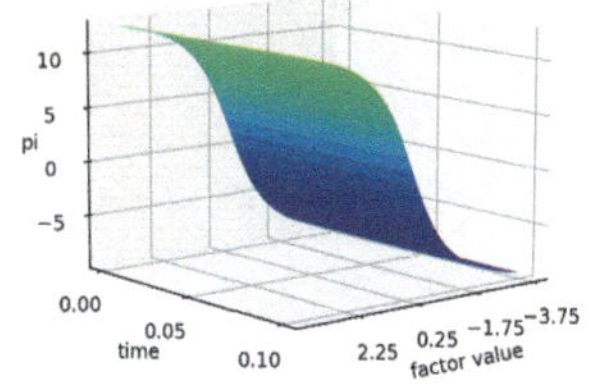

5 Conclusions

In this paper, we recall the theoretical results and numerical algorithms, that is PIA, obtained in [2] for an expected power utility maximization problem with consumption and investments. Applying the algorithm, we conduct a numerical experiment to solve the problem under a nonlinear stochastic factor model. Then, we verify that our PIA converges to the solution obtained by Hata et al. [1] relatively fast, and we also confirm that PIA obtained in [2] is applicable to a nonlinear stochastic factor model.

Acknowledgments Hiroaki Hata's research is supported by a Grant-in-Aid for Scientific Research (C), No. 20K11690, from Japan Society for the Promotion of Science. Kazuhiro Yasuda's research is supported by a Grant-in-Aid for Scientific Research (C), No. 21K03355, from Japan Society for the Promotion of Science.

References

1. Hata, H., Nagai, H., and Sheu, S.-J.: An optimal consumption problem for general factor models. SIAM J. Cont. Optim. **56**, 3149–3183 (2018)
2. Hata, H., Yasuda, K.: Policy improvement algorithm for an optimal consumption and investment problem under general stochastic factor models, submitted (2025)
3. Heath, D., Schweizer, M.: Martingales versus PDEs in finance: an equivalence result with examples. J. Appl. Probab. **37**, 947–957 (2000)
4. Howard, A.: Dynamic Programming and Markov Processes. MIT Press, Cambridge (1960)
5. Merton, R.C.: Optimal consumption and portfolio rules in a continuous time model. J. Econ. Theory **3**, 373–413 (1971)

An Extended Milstein Scheme for Effective Weak Approximation of Diffusions

Yuga Iguchi and Toshihiro Yamada

Abstract We propose a straightforward and effective method for discretizing multi-dimensional diffusion processes as an extension of Milstein scheme. The new scheme is explicitly given and can be simulated using Gaussian variates, requiring the same number of random variables as Euler-Maruyama (EM) scheme. We show that the proposed scheme has a weak convergence rate of one, which is consistent with other classical schemes like EM/Milstein schemes but involves fewer leading-order error terms. Due to the reduction of the error terms, the proposed scheme is expected to provide a more accurate estimation than alternative first-order schemes. We demonstrate that the weak error of the new scheme is effectively reduced compared with EM/Milstein schemes when the diffusion coefficients involve a small parameter. We conduct simulation studies on Asian option pricing in finance to showcase that our proposed scheme significantly outperforms EM/Milstein schemes, while interestingly, we find no differences in the performance between EM and Milstein schemes.

Keywords Asian option pricing · Euler-Maruyama scheme · Milstein scheme · Hypoelliptic diffusions · Stochastic differential equations · Weak approximation

1 Introduction

This work proposes a simple and effective weak approximation for Stochastic Differential Equations (SDEs) by extending *Milstein scheme*, one of the classical numerical schemes widely exploited in applications. Let $(\Omega, \mathcal{F}, \{\mathcal{F}_t\}_t, \mathbb{P})$ be a

Y. Iguchi
Department of Statistical Science, University College London, London, UK
e-mail: yuga.iguchi.21@ucl.ac.uk

T. Yamada (✉)
Graduate School of Economics, Hitotsubashi University, Tokyo, Japan
e-mail: toshihiro.yamada@r.hit-u.ac.jp

J. Akahori et al. (eds.), *Recent Developments in Stochastic Numerics and Computational Finance*, ICIAM2023 Springer Series 6,
https://doi.org/10.1007/978-981-95-0652-1_2

filtered probability space and $B_t = (B_t^1, \ldots, B_t^d)$, $t \geq 0$, be the d-dimensional standard Brownian motion defined upon the probability space. We consider the N-dimensional SDE specified as:

$$dX_t^x = b(X_t^x)dt + \sum_{j=1}^{d} \sigma_j(X_t^x)dB_t^j, \quad X_0^x = x \in \mathbb{R}^N, \tag{1}$$

with coefficients $b, \sigma_j : \mathbb{R}^N \to \mathbb{R}^N$, $1 \leq j \leq d$. Since the solution of SDEs is in general analytically unavailable, some time-discretisation is required to approximately compute the target dynamics. The accuracy of the approximation is usually measured by the weak/strong convergence: for a numerical scheme $\{\widetilde{X}_{kT/n}^{n,x}\}_{0 \leq k \leq n}$ starting from a point $x \in \mathbb{R}^N$ with a time $T > 0$ and a number of discretisation $n \in \mathbb{N}$,

$$\left|\mathbb{E}[f(X_T^x)] - \mathbb{E}[f(\widetilde{X}_T^{n,x})]\right| = \mathcal{O}(n^{-\alpha}); \tag{weak error}$$

$$\mathbb{E}\Big[\max_{0 \leq k \leq n} \left|X_{kT/n}^x - \widetilde{X}_{kT/n}^{n,x}\right|^2\Big] = \mathcal{O}(n^{-\beta}), \tag{strong error}$$

where $f : \mathbb{R}^N \to \mathbb{R}$ is some appropriate test function, and $\alpha, \beta > 0$ corresponds to the order of weak/strong convergence, respectively. For instance, it is well-known that for *Euler-Maruyama (EM) scheme*, $\alpha = 1.0$ and $\beta = 1.0$ (for instance see [10]).

As one of the popular discretisation schemes, the Milstein scheme has been investigated in many contexts of numerical analysis of SDEs. It achieves the same rate of weak convergence as the EM scheme ($\alpha = 1.0$) but has an improved rate of strong convergence, i.e., $\beta = 2.0$. The Milstein scheme is defined as: for $0 \leq k \leq n-1$,

$$\widetilde{X}_0^{\mathrm{Mil},n,x} = x;$$

$$\begin{aligned}\widetilde{X}_{(k+1)h}^{\mathrm{Mil},n,x} &= \widetilde{X}_{kh}^{\mathrm{Mil},n,x} + b(\widetilde{X}_{kh}^{\mathrm{Mil},n,x})h + \sum_{j=1}^{d} \sigma_j(\widetilde{X}_{kh}^{\mathrm{Mil},n,x})\Delta B_k^j \\ &\quad + \tfrac{1}{2} \sum_{j_1, j_2=1}^{d} g_{j_1 j_2}(\widetilde{X}_{kh}^{\mathrm{Mil},n,x})\big\{(\Delta B_k^{j_1} \Delta B_k^{j_2} - h \times \mathbf{1}_{j_1 = j_2}) - A_{j_1 j_2, k}\big\},\end{aligned} \tag{Milstein}$$

where we have set: $h = T/n$, $\Delta B_k^j = B_{(k+1)h}^j - B_{kh}^j$, $g_{j_1 j_2}(x) = \sum_{i=1}^{N} \sigma_{j_2}^i(x)\partial_i \sigma_{j_1}(x)$, $x \in \mathbb{R}^N$, and $A_{j_1 j_2, k}$ is the *Lévy area* defined as:

$$A_{j_1 j_2, k} = \int_{kh}^{(k+1)h} \int_{kh}^{s} dB_u^{j_1} dB_s^{j_2} - \int_{kh}^{(k+1)h} \int_{kh}^{s} dB_u^{j_2} dB_s^{j_1}. \tag{2}$$

The last term in (Milstein) is obtained from the higher order stochastic Taylor expansion of diffusion coefficients only and contributes to the improvement of the strong convergence, though it cannot be exactly simulated since the law of Lévy area $(A_{j_1 j_2,k})$ is intractable. The Milstein scheme is tractable only if the *commutative condition* holds, i.e., $g^i_{j_1 j_2}(x) - g^i_{j_2 j_1}(x) = 0,\ 1 \le i \le N,\ 1 \le j_1, j_2 \le d$ for any $x \in \mathbb{R}^N$. Precisely, under the commutative condition, the terms involving $A_{j_1 j_2,k}$ are cancelled out due to its anti-symmetric property. However, many important SDE models in applications, e.g., *stochastic volatility models* in financial engineering, do not satisfy the commutative condition. Thus, in practice, use of *truncated Milstein scheme*, i.e., the scheme (Milstein) with $A_{j_1 j_2,k}$ being replaced with 0, will be a one of realistic options. The scheme is defined as:

$$
\begin{aligned}
\widetilde{X}_0^{\mathrm{TMil},n,x} &= x; \\
\widetilde{X}_{(k+1)h}^{\mathrm{TMil},n,x} &= \widetilde{X}_{kh}^{\mathrm{TMil},n,x} + b(\widetilde{X}_{kh}^{\mathrm{TMil},n,x})h + \sum_{j=1}^{d} \sigma_j(\widetilde{X}_{kh}^{\mathrm{TMil},n,x})\Delta B_k^j \\
&\quad + \frac{1}{2} \sum_{j_1,j_2=1}^{d} g_{j_1 j_2}(\widetilde{X}_{kh}^{\mathrm{TMil},n,x})(\Delta B_k^{j_1} \Delta B_k^{j_2} - h \times \mathbf{1}_{j_1=j_2}).
\end{aligned}
\qquad \text{(T-Milstein)}
$$

The truncated scheme achieves the first order weak convergence, $\alpha = 1.0$. However, the removal of $A_{j_1 j_2,k}$ leads to deterioration of the strong convergence rate under the non-commutative condition and thus $\beta = 1.0$. Yamada [22] showed that for elliptic diffusions containing small parameters in both drift and diffusion coefficients, the truncated Milstein scheme can effectively reduce the weak error compared with the EM scheme. The truncated Milstein scheme is also exploited in the framework of multi-level Monte-Carlo (MLMC). For instance, [2] developed an antithetic multi-level Monte-Carlo based upon the truncated Milstein scheme.

One of the main focuses in this work is to investigate the (truncated) Milstein scheme from the perspective of weak approximation, motivated by the following question: *Can the Milstein scheme or the truncated Milstein scheme always produce a more accurate weak approximation than the EM scheme with the aid of higher order stochastic Taylor expansion of the diffusion coefficients?* The analytic result and numerical experiments provided later in this paper demonstrate that using the Milstein scheme does not necessarily have some advantages over the Euler-Maruyama scheme in weak approximation. In particular, this is shown by simulation studies in Sect. 4, where we find no significant differences in the performance of those two schemes. Then, instead of using these first order discretisation schemes, one might consider higher order weak approximation such as KLNV method [4, 5, 11, 14], Malliavin weight approach [6, 7, 13, 19, 21, 23], or weak second order sampling schemes [12, 20]. However, to improve the weak convergence, those schemes require extra computational efforts to take the Lévy area into account. For instance, most of the above methods require simulation of D-i.i.d. random variables for the local (one-step) transition, where D is larger than d, i.e., the dimension of driving Brownian motion. Thus, when n-times discretisation is introduced, the total number of generated random variables becomes nD, larger than nd required for EM and (truncated) Milstein schemes.

The central objective of this work is to propose a simple and effective first order weak approximation for a wide class of diffusions to outperform the Euler-Maruyama and the (truncated) Milstein scheme with a similar computational cost. In brief, the scheme we propose in this article features the following:

- It is always *explicit* even if the commutative condition does not hold.
- It extends scheme (T-Milstein) and involves additional terms from stochastic Taylor expansion of the drift coefficients.
- The total number of random variables to simulate the scheme is nd for n-times discretization, which is the same as EM and (truncated) Milstein schemes.

We refer to our proposed scheme as *extended Milstein scheme*. We will show (under some conditions on the SDE's coefficients) that the extended Milstein scheme achieves a first order weak approximation when the test function is bounded and measurable (non-smooth). In particular, we derive the analytic weak error expansion of the new scheme and compare it with that of EM scheme and (truncated) Milstein scheme. Then, for a class of hypo-elliptic diffusions with diffusion coefficients containing a small parameter $\varepsilon \in (0, 1)$, we quantitatively show that the discretization bias by the extended Milstein scheme is smaller than that of other first order schemes with the help of the small parameter ε. Finally, we present simulation studies showcasing that the extended Milstein scheme indeed significantly reduces the weak approximation bias compared with the other classical numerical schemes.

Notation We denote by $\mathscr{B}_b(\mathbb{R}^m)$, $m \geq 1$, the space of bounded and measurable functions $f : \mathbb{R}^m \to \mathbb{R}$. We write $C_b^\infty(\mathbb{R}^m; \mathbb{R}^n)$, $m, n \geq 1$ as the space of smooth functions $\varphi : \mathbb{R}^m \to \mathbb{R}^n$ with bounded derivatives of any order. We set $a = \sigma\sigma^\top$ and write the generator of the SDE (1) as:

$$\mathcal{L} \equiv \sum_{i=1}^{N} b^i(\cdot)\, \partial_i + \frac{1}{2} \sum_{i_1, i_2 = 1}^{N} a^{i_1 i_2}(\cdot)\, \partial_{i_1 i_2} \tag{3}$$

Throughout the paper, let $T > 0$, and we frequently use the notation of the test function $f \in \mathscr{B}_b(\mathbb{R}^N)$. We define the uniform norm $\| \cdot \|_\infty$ as: $\|f\|_\infty = \sup_{x \in \mathbb{R}^N} |f(x)|$ for $f \in \mathscr{B}_b(\mathbb{R}^N)$. We write $u(s, x) = \mathbb{E}[f(X_{T-s}^x)]$, $(s, x) \in [0, T] \times \mathbb{R}^N$ as the solution to the parabolic partial differential equation (PDE) specified as:

$$\begin{aligned} \partial_t u(t, x) + \mathcal{L}u(t, x) &= 0, \quad (t, x) \in [0, T) \times \mathbb{R}^N, \\ u(T, x) &= f(x). \end{aligned} \tag{4}$$

We define a function $\widetilde{b} : \mathbb{R}^N \to \mathbb{R}^N$ as:

$$\widetilde{b}^i(x) = b^i(x) - \frac{1}{2}\sum_{k=1}^{d}\sum_{j=1}^{N}\sigma_k^j(x)\partial_j\sigma_k^i(x), \quad x \in \mathbb{R}^N, \quad 1 \le i \le N, \tag{5}$$

which is the drift coefficient of the SDE when the Itô-type SDE (1) is written in the Stratonovich form. We identify the coefficients $\widetilde{b}, \sigma_j, \ 1 \le j \le d$ with the following vector fields:

$$\widetilde{L}_0 = \sum_{i=1}^{N}\widetilde{b}^i(\cdot)\,\partial_i, \quad L_j = \sum_{i=1}^{N}\sigma_j^i(\cdot)\,\partial_i, \quad 1 \le j \le d. \tag{6}$$

For two vector fields V, W, the Lie bracket is defined as:

$$[V, W] = VW - WV. \tag{7}$$

2 Preliminaries

We introduce two main conditions for the SDE (1), and then briefly review the work of [1], which provided an analytic result for weak approximation by Euler-Maruyama (EM) scheme under the above conditions. Based upon the analytic error expansion of EM scheme in the weak sense, we will compare the discretisation biases induced by the EM scheme and our new scheme proposed later in the next section.

2.1 *Conditions for the SDE*

We define sets of vector fields constructed from the Lie bracket (7) as:

$$\Sigma_0 = \{L_1, \ldots, L_d\},$$
$$\Sigma_m = \Big\{\{[\widetilde{L}_0, V], [L_j, V]\} : V \in \Sigma_{m-1}, 1 \le j \le d\Big\}, \quad m \ge 1.$$

For the SDE (1), or the Stratonovich-SDE with the drift (5), we assume the following two conditions under which [1] studied the weak error of Euler-Maruyama scheme when the test function is assumed to be bounded and measurable:

(H1) $b, \sigma_j, \ j = 1, \ldots, d$, are infinitely differentiable, and their derivatives of any order are bounded.

(H2) Uniform Hörmander's condition holds: there exists an integer $M \geq 0$ such that

$$\inf_{x \in \mathbb{R}^N} \inf_{\substack{\xi \in \mathbb{R}^N \\ \text{s.t.} \|\xi\|=1}} \sum_{0 \leq i \leq M} \sum_{V \in \Sigma_i} \langle V(x), \xi \rangle^2 > 0. \tag{8}$$

We note that under the condition (H2), the law of X_t^x is absolutely continuous with respect to (w.r.t.) the Lebesgue measure for any $x \in \mathbb{R}^N$ and $t > 0$. When (H2) holds with $M = 0$, the condition is typically interpreted as the uniformly elliptic condition, and thus the matrix $a := \sigma\sigma^\top$ is positive definite uniformly in the state $x \in \mathbb{R}^N$. Furthermore, together with the condition (H1), the condition (H2) leads to the existence of a smooth Lebesgue density of the law of X_t^x, see, e.g. [15].

2.2 *Weak Approximation by the Euler-Maruyama Scheme: Review*

We review the theoretical result for weak approximation by the EM scheme for hypo-elliptic diffusions. The EM scheme is defined as: for $x \in \mathbb{R}^N$, $n \geq 1$ and $h := T/n$,

$$\widetilde{X}_0^{\mathrm{EM},n,x} = x;$$

$$\widetilde{X}_{(k+1)h}^{\mathrm{EM},n,x} = \widetilde{X}_{kh}^{\mathrm{EM},n,x} + b(\widetilde{X}_{kh}^{\mathrm{EM},n,x})h + \sum_{j=1}^{d} \sigma(\widetilde{X}_{kh}^{\mathrm{EM},n,x})\,\Delta B_k^j, \quad 0 \leq k \leq n-1, \tag{EM}$$

Under conditions (H1)–(H2), [1] showed that the EM scheme achieves the first order weak convergence:

Theorem 1 (Bally and Talay [1]) *Let $T > 0$, $x \in \mathbb{R}^N$ and $f \in \mathscr{B}_b(\mathbb{R}^N)$. Under the conditions (H1)–(H2), it holds that:*

$$\mathbb{E}[f(X_T^x)] - \mathbb{E}[f(\widetilde{X}_T^{\mathrm{EM},n,x})] = \frac{T}{n}\, C_f^{\mathrm{EM}}(T,x) + \frac{1}{n^2}\, R_f^{\mathrm{EM}}(T,x), \tag{9}$$

with the remainder terms $C_f^{\mathrm{EM}}(T,x)$ and $R_f^{\mathrm{EM}}(T,x)$ specified as follows: the leading order term is given by $\frac{T}{n}C_f^{\mathrm{EM}}(T,x)$ with

$$C_f^{\mathrm{EM}}(T,x) = \int_0^T \mathbb{E}[\Phi^{\mathrm{EM}}(s, X_s^x)]ds,$$

where $\Phi^{\mathrm{EM}} : [0,\infty) \times \mathbb{R}^N \to \mathbb{R}$ *is defined as:*

$$\Phi^{\mathrm{EM}}(t,x) = -\tfrac{1}{2}\sum_{i,j=1}^{N} b^i(x)b^j(x)\partial_{ij}u(t,x) - \tfrac{1}{2}\sum_{i,j,k=1}^{N} b^i(x)a^{jk}(x)\partial_{ijk}u(t,x)$$

$$-\tfrac{1}{8}\sum_{i,j,k,l=1}^{N} a^{ij}(x)a^{kl}(x)\partial_{ijkl}u(t,x) - \tfrac{1}{2}\partial_t^2 u(t,x)$$

$$-\sum_{i=1}^{N} b^i(x)\partial_t\partial_i u(t,x) - \tfrac{1}{2}\sum_{i,j=1}^{N} a^{ij}(x)\partial_t\partial_{ij}u(t,x). \tag{10}$$

Also, there exist real numbers $q, Q > 0$ *and a non-decreasing function* $K(\cdot)$ *such that*

$$|C_f^{\mathrm{EM}}(T,x)| + |R_f^{\mathrm{EM}}(T,x)| \le K(T)\|f\|_\infty \frac{1+|x|^Q}{T^q}. \tag{11}$$

Our objective is to compare the weak approximation error by EM scheme with that by the extended Milstein scheme proposed later in (18). To this end, we here provide an expression for Φ^{EM} without requiring the partial derivatives of u w.r.t. the time variable. For simplicity of the notation, we write:

$$L_0 \equiv \mathcal{L} = \sum_{i=1}^{N} b^i(\cdot)\partial_i + \tfrac{1}{2}\sum_{1\le i,j\le N} a^{ij}(\cdot)\partial_{ij}. \tag{12}$$

Making use of $\partial_t u = -\mathcal{L}u$, we obtain the following result whose proof is postponed to Appendix 1: Proof of Lemma 1.

Lemma 1 *The function* $\Phi^{\mathrm{EM}} : [0,\infty) \times \mathbb{R}^N \to \mathbb{R}$ *is given as:*

$$\Phi^{\mathrm{EM}}(s,x) = \Phi_1(s,x) + \Phi_2(s,x) + \Phi_3(s,x), \quad (s,x) \in [0,\infty) \times \mathbb{R}^N, \tag{13}$$

with

$$\Phi_1(s,x) \equiv \tfrac{1}{2}\sum_{i=1}^{N} L_0 b^i(x)\partial_i u(s,x)$$

$$+\tfrac{1}{2}\sum_{i,j=1}^{N}\sum_{m=1}^{d} \sigma_m^i(x)\{L_m b^j(x) + L_0\sigma_m^j(x)\}\partial_{ij}u(s,x), \tag{14}$$

$$\Phi_2(s,x) \equiv \tfrac{1}{2}\sum_{i,j,k}^{N}\sum_{m_1,m_2=1}^{d} L_{m_1}\sigma_{m_2}^i(x)\sigma_{m_1}^j(x)\sigma_{m_2}^k(x)\partial_{ijk}u(s,x)$$
$$+\tfrac{1}{8}\sum_{i,j=1}^{N}\sum_{m_1,m_2=1}^{d} L_{m_1}\sigma_{m_2}^i(x)\big\{L_{m_1}\sigma_{m_2}^j(x)+L_{m_2}\sigma_{m_1}^j(x)\big\}\partial_{ij}u(s,x), \tag{15}$$

$$\Phi_3(s,x) \equiv \tfrac{1}{8}\sum_{i,j=1}^{N}\sum_{m_1,m_2=1}^{d} L_{m_1}\sigma_{m_2}^i(x)\,[L_{m_1},L_{m_2}]^j(x)\,\partial_{ij}u(s,x). \tag{16}$$

Thus, due to Theorem 1 and Lemma 1, the weak approximation error invoked by the EM scheme involves leading order term of size $\mathcal{O}(T/n)$ that is specified as:

$$\frac{T}{n}\times C_f^{\mathrm{EM}}(T,x)=\frac{T}{n}\times\int_0^T \mathbb{E}\big[\Phi_1(s,x)+\Phi_2(s,x)+\Phi_3(s,x)\big]ds. \tag{17}$$

3 Extended Milstein Scheme and Weak Error Analysis

In this section, we propose the extended Milstein scheme mentioned in Sect. 1 so that it produces a more effective weak approximation than other classical numerical schemes. We then show the analytic weak error expansions for the EM, the (truncated) Milstein and the extended Milstein scheme in the manner of Theorem 1 as the core result in this paper. In the last subsection, we apply the main result for a class of hypo-elliptic diffusions with a small parameter in the diffusion coefficients to clarify an advantage of the proposed scheme over other classical schemes.

3.1 Extended Milstein Scheme

We introduce a new discretisation scheme as an extension of the truncated Milstein scheme. For simplicity of notation, we write $b \equiv \sigma_0$. Let $T > 0$, $x \in \mathbb{R}^N$ and $h = T/n$ with $n \in \mathbb{N}$. Then, we propose the extended Milstein scheme as follows: for $0 \le k \le n-1$,

$$\bar{X}_0^{n,x} = x;$$

$$\bar{X}_{(k+1)h}^{n,x} = \bar{X}_{kh}^{n,x} + \sum_{j=0}^{d}\sigma_j(\bar{X}_{kh}^{n,x})\Delta B_k^j$$
$$+\sum_{0\le j_1,j_2\le d} L_{j_1}\sigma_{j_2}(\bar{X}_{kh}^{n,x})\times\frac{1}{2}\Big\{\Delta B_k^{j_1}\Delta B_k^{j_2}-h\times\mathbf{1}_{j_1=j_2\neq 0}\Big\}, \tag{18}$$

where we interpret $\Delta B_k^0 = h$. We notice that the scheme (18) is explicitly given and simulated by d-dimensional Brownian increments only for each iteration. Also, it includes terms from the stochastic Taylor expansion of the drift function σ_0 and $L_0\sigma_{j_2}$.

We have the following result on the proposed scheme (18):

Theorem 2 *Let $T > 0$, $x \in \mathbb{R}^N$ and $f \in \mathscr{B}_b(\mathbb{R}^N)$. Under the conditions (H1)–(H2), it holds that:*

$$\mathbb{E}[f(X_T^x)] - \mathbb{E}[f(\bar{X}_T^{n,x})] = \frac{T}{n} C_f(T,x) + \frac{1}{n^2} R_f(T,x), \tag{19}$$

with the terms $C_f(T,x)$ and $R_f(T,x)$ specified as follows: the first term of the right hand side of (19) is the leading order term with

$$C_f(T,x) \equiv \int_0^T \mathbb{E}\big[\Phi_3(s, X_s^x)\big]ds, \tag{20}$$

where Φ_3 is given in (16). Also, there exist real numbers $q, Q > 0$ and a non-decreasing function $K(\cdot)$ such that

$$|C_f(T,x)| + |R_f(T,x)| \le K(T)\|f\|_\infty \frac{1+|x|^Q}{T^q}. \tag{21}$$

The proof of Theorem 2 is contained in Sect. 5. Notice that the leading error term by the extended Milstein scheme (18) is given as:

$$\frac{T}{n} C_f(T,x) = \frac{T}{n}\int_0^T \mathbb{E}[\Phi_3(s,x)]ds. \tag{22}$$

Upon consideration of Theorem 1, both the Euler-Maruyama scheme and the extended Milstein scheme share the same rate of weak convergence $\mathcal{O}(n^{-1})$, but the extended Milstein scheme (18) invoke the fewer error terms than Euler-Maruyama scheme does: See $\frac{T}{n} C_f^{\mathrm{EM}}(T,x)$ and $\frac{T}{n} C_f(T,x)$ given in (17) and (22), respectively.

Remark 1 It is shown that the leading weak error terms of the (intractable) Milstein scheme and the (tractable) truncated Milstein scheme are given as:

$$\tfrac{T}{n}\int_0^T \mathbb{E}\big[\Phi_1(s, X_s^x)\big]ds, \qquad \tfrac{T}{n}\int_0^T \mathbb{E}\big[\Phi_1(s, X_s^x) + \Phi_3(s, X_s^x)\big]ds, \tag{23}$$

respectively. The above error terms and Theorem 2 imply that the appearance of the term Φ_3 results from removing the Lévy area in the definition of the truncated Milstein/extended Milstein scheme. Furthermore, in the case of the proposed scheme (18), the weak error term $\frac{T}{n}\int_0^T \mathbb{E}[\Phi_1(s,x)]ds$ does not appear. This is because the scheme (18) involves the following terms in its definition:

$$L_0\sigma_0(\bar{X}^{n,x}_{kh}) \times \frac{h^2}{2},$$

$$\sum_{1\le j\le d} \left\{L_0\sigma_j(\bar{X}^{n,x}_{kh}) + L_j\sigma_0(\bar{X}^{n,x}_{kh})\right\} \times \frac{1}{2}\Delta B^j_k h, \quad 0 \le k \le n-1. \tag{24}$$

3.2 Hypo-Elliptic Diffusions with a Small Diffusion Parameter

Based on the main result (Theorem 2), we analytically demonstrate the advantage of the proposed scheme (18) over the Euler-Maruyama/Milstein schemes in terms of weak approximation. In particular, when the diffusion coefficients involve a small parameter $\varepsilon \in (0, 1)$, which often appears in applications, we quantitatively show that the leading weak error term of the extended scheme (18) is smaller than that of EM and (truncated) Milstein scheme by incorporating ε into the error bound. To observe this, we introduce the following hypo-elliptic diffusion:

$$\begin{aligned} dX^{x,\varepsilon}_t = \begin{bmatrix} dX^{x,\varepsilon}_{R,t} \\ dX^{x,\varepsilon}_{S,t} \end{bmatrix} &= \begin{bmatrix} b_R(X^{x,\varepsilon}_t) \\ b_S(X^{x,\varepsilon}_t) \end{bmatrix} dt \\ &+ \sum_{j=1}^{d} \begin{bmatrix} \epsilon\,\sigma_j(X^{x,\varepsilon}_t) \\ \mathbf{0}_{N_S} \end{bmatrix} dB^j_t, \quad X^{x,\varepsilon}_0 = (X^{x,\varepsilon}_{R,0}, X^{x,\varepsilon}_{S,0}) = (x_R, x_s) \in \mathbb{R}^N, \end{aligned} \tag{25}$$

where $\varepsilon \in (0, 1)$ and $X^{x,\varepsilon}_{R,t} \in \mathbb{R}^{N_R}$, $X^{x,\varepsilon}_{S,t} \in \mathbb{R}^{N_S}$ with integers $N_R, N_S \ge 1$ such that $N_R + N_S = N$. In the above, the coefficients are specified as:

$$b_R : \mathbb{R}^N \to \mathbb{R}^{N_R}, \quad b_S : \mathbb{R}^N \to \mathbb{R}^{N_S}, \quad \sigma_j : \mathbb{R}^N \to \mathbb{R}^{N_R}, \quad 1 \le j \le d.$$

When considering the model (25), we write $u^\varepsilon(s, x)$, $(s, x) \in (0, \infty) \times \mathbb{R}^d$ instead of $u(s, x)$ to emphasise the dependence of the parameter ε. We then introduce the following result whose proof is provided in Appendix 3: Proof of Proposition 1.

Proposition 1 *Let $T > 0$, $\alpha \in \{1, \ldots, N\}^k$, $k \in \mathbb{N}$ and $g \in \mathbb{R}^N \to \mathbb{R}$ be a smooth function with polynomial growth. Work under the conditions (H1) and (H2) with $M = 1$. Then, there exist a non-decreasing function $K(\cdot)$ and constants $q, Q > 0$ independent of $\varepsilon \in (0, 1)$, n, N such that*

$$\left| \int_0^T \mathbb{E}\big[g(X^{x,\varepsilon}_s)\partial_\alpha u^\varepsilon(s, X^{x,\varepsilon}_s)\big]ds \right| \le K(T)\frac{\|f\|_\infty}{T^q \varepsilon^k} \times (1 + |x|^Q). \tag{26}$$

Table 1 Comparison of first order schemes

Scheme	Leading error term	Requirement of Lévy area
EM	$\frac{T}{n}\sum_{1\le i\le 3}\int_0^T \mathbb{E}[\Phi_i(s, X_s^x)]ds$	No
Milstein	$\frac{T}{n}\int_0^T \mathbb{E}[\Phi_1(s, X_s^x)]ds$	Yes
T-Milstein	$\frac{T}{n}\int_0^T \mathbb{E}[\Phi_1(s, X_s^x)+\Phi_3(s, X_s^x)]ds$	No
New (18)	$\frac{T}{n}\int_0^T \mathbb{E}[\Phi_3(s, X_s^x)]ds$	No

We also adjust the notation of the EM/Milstein/truncated Milstein/extended Milstein scheme applied to the SDE (25) as $\widetilde{X}_{kh}^{\mathrm{EM},n,x,\varepsilon}/\widetilde{X}_{kh}^{\mathrm{Mil},n,x,\varepsilon}/\widetilde{X}_{kh}^{\mathrm{TMil},n,x,\varepsilon}/\bar{X}_{kh}^{n,x,\varepsilon}$, respectively. Application of Proposition 1 to the leading error terms for the first order schemes presented in Table 1, together with Theorems 1 and 2, leads to the following result.

Corollary 1 *Let $T > 0$, $x \in \mathbb{R}^N$, $\varepsilon \in (0, 1)$ and $n \in \mathbb{N}$. Assume conditions (H1) and (H2) with $M = 1$ hold.*

1. *Let $f_{\mathrm{Lip}} : \mathbb{R}^N \to \mathbb{R}$ be Lipschitz continuous. There exist positive constants C_1^w, q_1^w, $w \in \{\mathrm{EM, TMil, Mil, New}\}$, independent of n, ε and x such that*

$$\left|\mathbb{E}\big[f_{\mathrm{Lip}}(X_T^{x,\varepsilon})\big] - \mathbb{E}\big[f_{\mathrm{Lip}}(\widetilde{X}_T^{w,n,x,\varepsilon})\big]\right| \le C_1^w \times \frac{\|\nabla f_{\mathrm{Lip}}\|_\infty}{n}(1+|x|^{q_1^w}),$$
$$w \in \{\mathrm{EM, TMil, Mil}\};$$

$$\left|\mathbb{E}\big[f_{\mathrm{Lip}}(X_T^{x,\varepsilon})\big] - \mathbb{E}\big[f_{\mathrm{Lip}}(\bar{X}_T^{n,x,\varepsilon})\big]\right| \le C_1^{\mathrm{New}}\varepsilon^3 \times \frac{\|\nabla f_{\mathrm{Lip}}\|_\infty}{n}(1+|x|^{q_1^{\mathrm{New}}}).$$

2. *Let $f \in \mathscr{B}_b(\mathbb{R}^N)$. There exist positive constants C_2^w, q_2^w, $w \in \{\mathrm{EM, TMil, Mil, New}\}$, independent of n, ε and x such that*

$$\left|\mathbb{E}\big[f(X_T^{x,\varepsilon})\big] - \mathbb{E}\big[f(\widetilde{X}_T^{w,n,x,\varepsilon})\big]\right| \le C_2^w \times \frac{\|f\|_\infty}{n\varepsilon}(1+|x|^{q_2^w}),$$
$$w \in \{\mathrm{EM, TMil, Mil}\};$$

$$\left|\mathbb{E}\big[f(X_T^{x,\varepsilon})\big] - \mathbb{E}\big[f(\bar{X}_T^{n,x,\varepsilon})\big]\right| \le C_2^{\mathrm{New}}\varepsilon^2 \times \frac{\|f\|_\infty}{n}(1+|x|^{q_2^{\mathrm{New}}}).$$

Corollary 1 demonstrates that the extended scheme (18) has a better upper bound with the help of $\varepsilon \in (0, 1)$, compared with other schemes. The differences in the upper bounds result from that EM, truncated Milstein and Milstein scheme involve the term $\int_0^T \mathbb{E}\big[\Phi_1(s, X_s^{x,\varepsilon})\big]ds$ in the leading error, while the term is eliminated in the case of the extended Milstein scheme (see also Remark 1). Furthermore, due to the presence of ε^{α}, $\alpha > 0$ in the bound, the proposed scheme can behave nearly as a second order weak approximation up to some number of discretisation $n \in \mathbb{N}$ given a sufficiently small $\varepsilon \in (0, 1)$. Thus the extended Milstein scheme provides an asymptotic approximation which is slightly different to the discretisation schemes or expansions in [7, 18, 19]. This is also observed in the simulation studies in the following section.

Remark 2 The interpretation of Corollary 1 is as follows. When the SDE is driven by a small noise, the model is mainly dominated by the drift function and thus, it is critical to include the higher order expansion terms of the drift function to increase the precision of approximation. We iterate here that the three schemes (EM/truncated Milstein/Milstein) do not exploit such higher order terms from the drift function while the extended Milstein scheme does.

4 Simulation Studies

4.1 Asian Call Option Pricing with Black-Scholes Model

We consider the following 2-dimensional hypo-elliptic SDE:

$$\begin{aligned} dS_t &= rS_t dt + \sigma S_t dB_t^1; \\ dA_t &= S_t dt, \end{aligned} \tag{27}$$

with the parameters $r > 0$ and $\sigma > 0$. S_t and A_t represent the price of the underlying asset and the time accumulation of the asset value, respectively. Then, the price of *Asian call option* with the strike price K and the maturity T is formulated as $D \times \mathbb{E}\big[\varphi_K(A_T/T)\big]$, where the text function is $\varphi_K(x) = \max(x - K, 0)$ for $x \in \mathbb{R}^N$, and $D = e^{-rT}$ is the discount factor. We note that for the SDE (27) the commutative condition holds and thus the Milstein scheme coincides with the truncated Milstein scheme. We consider the following setting for parameter values: $r = 0.1$, $\sigma \in \{0.4, 0.8\}$, $T = 1.0$ and $(S_0, A_0) = (100.0, 0.0)$. We compute the benchmark value by the standard Monte-Carlo (MC) method using the EM scheme with the number of paths $M = 10^7$ and discretization $n = 2^{10}$. We compare the performances of three numerical schemes, EM, Milstein and the extended Milstein scheme via the following MC estimates: for a number of MC paths M and discretisation n,

$$f_K^{w}(M, n) = \tfrac{1}{M} \sum_{j=1}^{M} D \times \varphi_K\big(\widetilde{A}_T^{w,n,[j]}/T\big), \qquad w \in \{\text{EM}, \text{Mil}, \text{New}\},$$

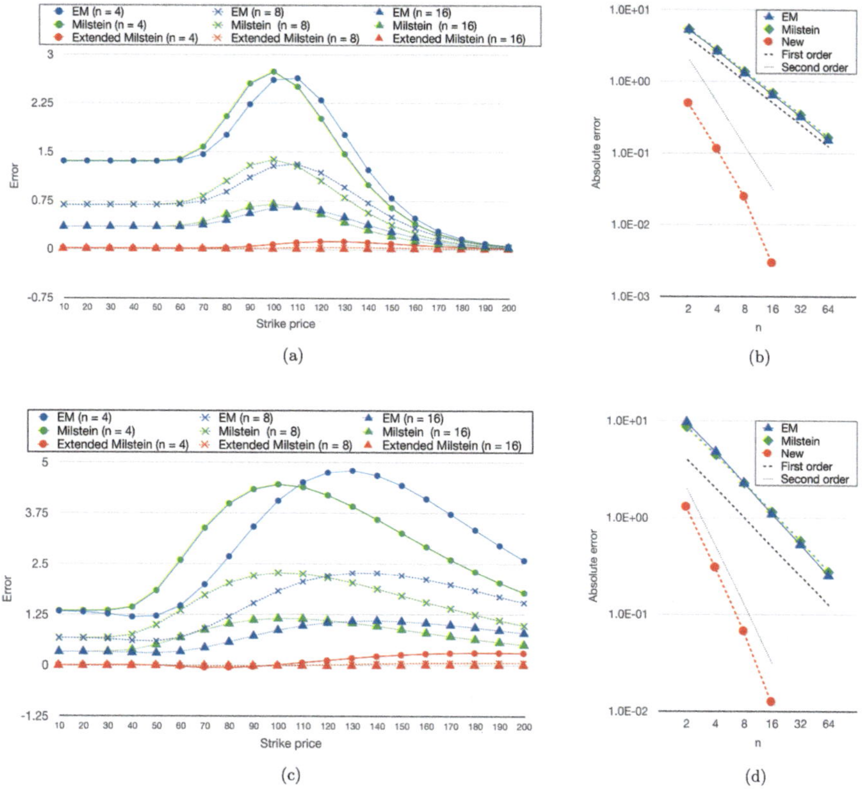

Fig. 1 Asian call option pricing (BS model). (**a**) Error of estimation ($\sigma = 0.4$). (**b**) Weak convergence ($\sigma = 0.4$). (**c**) Error of estimation ($\sigma = 0.8$). (**d**) Weak convergence ($\sigma = 0.8$)

where $\widetilde{A}_T^{\mathrm{EM},n,[j]}$, $\widetilde{A}_T^{\mathrm{Mil},n,[j]}$, $\widetilde{A}_T^{\mathrm{New},n,[j]}$ are the j-th trajectory of the EM scheme, the truncated Milstein scheme and the extended Milstein scheme (18) applied to the model (27), respectively. In Fig. 1a, c, we plot

$$\mathscr{E}_K^w(M,n) \equiv (\text{Benchmark value})_K - f_K^w(M,n), \qquad w \in \{\text{EM}, \text{Mil}, \text{New}\},$$

with $M = 10^6$ for $n = 4, 8, 16$ and $K = 10, 20, \ldots, 200$, where we applied Quasi-Monte-Carlo (QMC) for computing $f_K^w(M,n)$. We observe that the proposed scheme provides very accurate estimates for all the strike prices and outperforms the EM/Milstein schemes. Also, Figure 1b, d show

$$\sup_{K \in \{10,20,\ldots,200\}} \left|\mathscr{E}_K^w(10^6, n)\right|, \qquad w \in \{\text{EM}, \text{Mil}, \text{New}\}, \tag{28}$$

for various numbers of discretisation n. It is noteworthy that there is no significant difference between the performance of the EM and Milstein schemes.

4.2 Asian Digital Option Pricing with Heston Model

We consider the following 3-dimensional SDE:

$$\begin{aligned} dS_t &= \sqrt{\sigma_t} S_t dB_t^1; \\ d\sigma_t &= \alpha(\theta - \sigma_t)dt + \nu\sqrt{\sigma_t}(\rho dB_t^1 + \sqrt{1-\rho^2} dB_t^2); \\ dA_t &= S_t dt, \end{aligned} \tag{29}$$

with parameters $\alpha, \nu, \theta > 0$ and $\rho \in [-1, 1]$ satisfying $2\alpha\theta > \nu^2$ so that the process $\{\sigma_t\}_{t\geq 0}$ is strictly positive. The pair $\{S_t, \sigma_t\}_{t\geq 0}$ represents the process of the underlying asset and its volatility, and is a famous stochastic volatility model called *Heston model* in the context of financial mathematics. We then estimate the price of Asian digital option with the coupon $\mathrm{Cpn} > 0$, the strike price $K > 0$ and the maturity $T > 0$, given as:

$$\mathrm{Cpn} \times \mathbb{E}\big[H_K(A_T/T)\big], \tag{30}$$

where $H_K(x) \equiv \mathbf{1}_{x\geq K}$. Note that the test function is non-smooth. Since (30) does not admit a closed-form solution, we apply the quasi-Monte Carlo method accompanied by some discretisation scheme to approximate the quantity. In particular, the commutative condition does not hold for the model (29), and then the standard Milstein scheme is intractable due to the presence of Lévy area. We thus compare the performance of these three numerical schemes via the following Monte-Carlo estimates:

$$f_K^w(M, n) = \frac{1}{M}\sum_{j=1}^{M} \mathrm{Cpn} \times H_K\big(\widetilde{A}_T^{w,n,[j]}/T\big), \qquad w \in \{\mathrm{EM}, \mathrm{TMil}, \mathrm{New}\}$$

with the number of trajectories M and the number of discretisation n, where $\widetilde{A}_T^{\mathrm{EM},n,[j]}$, $\widetilde{A}_T^{\mathrm{TMil},n,[j]}$, $\widetilde{A}_T^{\mathrm{New},n,[j]}$ are the j-th trajectory of the EM scheme, the truncated Milstein scheme and the extended Milstein scheme (18) applied to the model (29), respectively. We set the parameter values as: $\alpha = 2.0$, $\theta = 0.09$, $\nu = 0.1$, $\rho = 0.7$, $T = 1.0$, $(S_0, \sigma_0, A_0) = (100, 0.09, 0.0)$ and $\mathrm{Cpn} = 100$. We estimate the benchmark value by applying the standard Monte-Carlo method with the EM scheme as: $(\text{Benchmark value})_K = f_K^{\mathrm{EM}}(10^7, 2^{11})$. In Fig. 2a, we plot

$$\mathscr{E}_K^w(M, n) \equiv (\text{Benchmark value})_K - f_K^w(M, n), \qquad w \in \{\mathrm{EM}, \mathrm{TMil}, \mathrm{New}\},$$

with $M = 10^6$ for $n = 4, 8, 16$ and $K = 10, 20, \ldots, 200$, where we applied Quasi-Monte-Carlo (QMC) for computing $f_K^w(M, n)$. Also, Fig. 2b indicates

$$\sup_{K\in\{10,20,\ldots,200\}} \big|\mathscr{E}_K^w(10^6, n)\big|, \qquad w \in \{\mathrm{EM}, \mathrm{TMil}, \mathrm{New}\}, \tag{31}$$

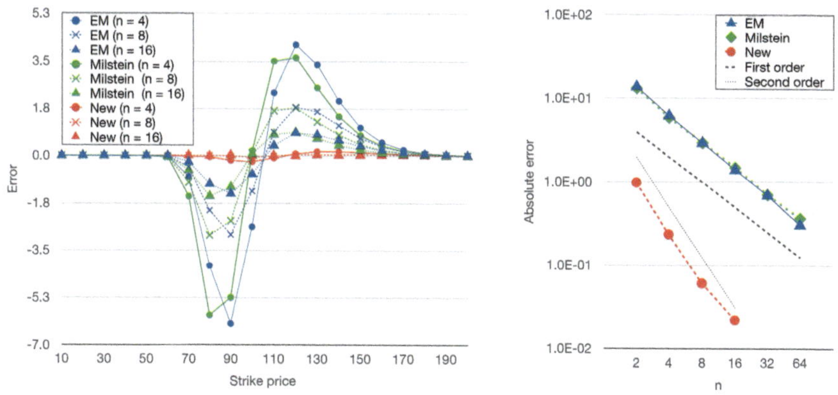

(a) Error of estimation.

(b) Weak convergence.

Fig. 2 Asian digital option pricing (Heston model) with three numerical schemes. (**a**) Error of estimation. (**b**) Weak convergence

for various numbers of discretization n. As we observed in the previous numerical experiments on the Asian BS model, the proposed scheme again provides quite accurate estimates compared with the EM scheme and the truncated Milstein scheme. Notably, from Fig. 2b, the bias induced by the proposed scheme is less than a tenth of that by the EM scheme or the truncated Milstein scheme, for $n \in \{2, 4, 8, 16\}$ and achieves nearly the second order weak convergence, which is expected from the argument in Sect. 3.2 for small diffusions.

5 Proof of Theorem 2

Most of our proof relies on the argument in [1] that shows Theorem 1, i.e., weak convergence of EM scheme under non-smooth test functions. The strategy is naturally applied to the case of extended Milstein scheme (18), and our main focus is to derive an explicit form of the leading error term of weak approximation by the extended Milstein scheme, which is associated with the local error expansion by the scheme (Proposition 2). The technical results introduced later (Lemmas 2, 3 and 4) are also shown by following the proofs in [1] and adjusting the argument for EM scheme with our proposed scheme (18). Thus, the detailed proof of these results will be omitted.

We introduce some notation for the proof. Let $T > 0$, $h = T/n$, $n \geq 1$ and we assume $f \in \mathscr{B}_b(\mathbb{R}^N)$. We introduce families of Markov operators $\{P_t\}_{t\geq 0}$ and $\{Q_t\}_{t\geq 0}$ as:

$$P_t f(x) = \mathbb{E}[f(X_t^x)], \quad Q_t f(x) = \mathbb{E}[f(\bar{X}_t^x)], \quad t > 0,\ x \in \mathbb{R}^N, \tag{32}$$

with $\bar{X}_t^x$ being the one step extended Milstein scheme specified as:

$$\bar{X}_t^x = x + \sum_{j=0}^{d} \sigma_j(x) B_t^j + \sum_{0 \le j_1, j_2 \le d} L_{j_1}\sigma_{j_2}(x) \times \tfrac{1}{2}(B_t^{j_1} B_t^{j_2} - t\mathbf{1}_{j_1=j_2\neq 0}), \tag{33}$$

where we recall $\sigma_0 = b$ and $L_0 = \mathcal{L}$ given in (3). We note that k-times application of the Markov operator Q_h gives:

$$(Q_h)^k \varphi(x) = \mathbb{E}[\varphi(\bar{X}_{kh}^{n,x})], \quad \varphi \in \mathscr{B}_b(\mathbb{R}^N), \tag{34}$$

where $\bar{X}_{kh}^{n,x}$ is the Markov chain of the one step extended Milstein scheme defined as (18). First, we have the following result associated with the local weak approximation by the extended Milstein scheme (18):

Proposition 2 *Let $x \in \mathbb{R}^N$ and $h = T/n,\ T > 0,\ n \in \mathbb{N}$. We have*

$$(Q_h)^k (P_h - Q_h) P_{T-(k+1)h} f(x) = h^2\, \mathbb{E}\Big[\Phi_3\big(kh, \bar{X}_{kh}^{n,x}\big)\Big] + \widetilde{\Phi}_k^{(n)}(h, x),$$
$$0 \le k \le n-2, \tag{35}$$

where Φ_3 is defined as (14), and $\widetilde{\Phi}_k^{(n)}(h, x)$ is determined from the sum of terms given in the form of

$$\mathbb{E}\left[e_\alpha(\bar{X}_{kh}^{n,x}) \int_{kh}^{(k+1)h} \int_{kh}^{s_1} \int_{kh}^{s_2} g_\alpha(\bar{X}_{s_3}^{n,x}) \partial_\alpha u(s_3, \bar{X}_{s_3}^{n,x}) ds_3 ds_2 ds_1 \right], \tag{36}$$

where $\alpha \in \{1, \ldots, N\}^\ell$ with $\ell \le 6$, and the functions e_α, g_α are given as products of partial derivatives of coefficients σ_j^i, $0 \le j \le d$, $1 \le i \le N$.

The proof of Proposition 2 is provided in Appendix 2: Proof of Proposition 2. Proposition 2 gives the global weak error expansion for the proposed scheme (18) in the following form:

$$P_T f(x) - (Q_h)^n f(x) = \sum_{k=0}^{n-1} (Q_h)^k (P_h - Q_h) P_{T-(k+1)h} f(x)$$
$$= h^2 \sum_{k=0}^{n-2} \mathbb{E}[\Phi_3(kh, \bar{X}_{kh}^{n,x})] + \sum_{k=0}^{n-2} \widetilde{\Phi}_k^{(n)}(h, x) + (Q_h)^{n-1}(P_h - Q_h) f(x) \tag{37}$$

$$= h \int_0^T \mathbb{E}[\Phi_3(s, X_s^x)] ds + \sum_{i=1}^{4} \mathcal{E}_{i,n}^f(h, x), \tag{38}$$

where we have set:

$$\mathcal{E}_{1,n}^{f}(h,x) \equiv h^2 \sum_{k=0}^{n-2} \mathbb{E}\big[\Phi_3(kh, X_{kh}^x)\big] - h \int_0^T \mathbb{E}\big[\Phi_3(s, X_s^x)\big] ds;$$

$$\mathcal{E}_{2,n}^{f}(h,x) \equiv h^2 \sum_{k=0}^{n-2} \mathbb{E}\big[\Phi_3(kh, \bar{X}_{kh}^{n,x})\big] - h^2 \sum_{k=0}^{n-2} \mathbb{E}\big[\Phi_3(kh, X_{kh}^x)\big];$$

$$\mathcal{E}_{3,n}^{f}(h,x) \equiv \sum_{k=0}^{n-2} \widetilde{\Phi}_k^{(n)}(h,x), \quad \mathcal{E}_{4,n}^{f}(h,x) \equiv (Q_h)^{n-1}(P_h - Q_h) f(x).$$

To show the upper bounds of the terms $\mathcal{E}_{i,n}^{f}(h,x)$, $1 \le k \le 4$, we introduce the following three lemmas (corresponding results based upon the EM scheme are found in [1] as Lemmas 4.1, 4.2 and 4.3, respectively):

Lemma 2 *For any multi-index $\alpha \in \{1, \dots, N\}^{\ell}$, $\ell \in \mathbb{N}$ and any smooth function $g : \mathbb{R}^N \to \mathbb{R}$ with the polynomial growth, there exist a non-decreasing function $K(\cdot)$ and constants q, $Q > 0$ independent of n, T such that*

$$\big|\mathbb{E}\big[g(X_s^x)\partial_\alpha P_{T-s} f(X_s^x)\big]\big| \le K(T)\frac{\|f\|_\infty}{T^q}(1 + |x|^Q) \tag{39}$$

for all $s \in [0, T]$ and

$$\big|\mathbb{E}\big[g(\bar{X}_s^{n,x})\partial_\alpha P_{T-s} f(\bar{X}_s^{n,x})\big]\big| \le K(T)\frac{\|f\|_\infty}{T^q}(1 + |x|^Q) \tag{40}$$

for all $s \in [0, T - T/n]$.

Lemma 3 *Let $\alpha \in \{1, \dots, N\}^{l_1}$, $\beta \in \{1, \dots, N\}^{l_2}$, $l_1, l_2 \in \mathbb{N}$ and $g, \widetilde{g} : \mathbb{R}^N \to \mathbb{R}$ be smooth functions with polynomial growth. Set $\varphi_s : \mathbb{R}^N \to \mathbb{R}$, $s \in [0, T]$ as: $y \mapsto \varphi_s(y) \equiv \widetilde{g}(y)\partial_\beta P_{T-s} f(y)$. Then, there exist a non-decreasing function $K(\cdot)$ and constants q, $Q > 0$ independent of n, T such that for all $s \in [0, T - T/n]$ and for all $t \in [0, s - T/n]$,*

$$\big|\mathbb{E}[g(\bar{X}_t^{n,x})\partial_\alpha P_{s-t}\varphi_s(\bar{X}_t^{n,x})]\big| \le K(T)\frac{\|f\|_\infty}{T^q}(1 + |x|^Q). \tag{41}$$

Lemma 4 *There exist a non-decreasing function $K(\cdot)$ and a constant $Q > 0$ independent of n, T such that*

$$\big|\mathcal{E}_{4,n}^{f}(h,x)\big| \le K(T)\frac{\|f\|_\infty}{n^2}(1 + |x|^Q). \tag{42}$$

Proof of Lemmas 2, 3 and 4 These are shown via the similar argument used to prove Lemmas 4.1, 4.2, 4.3 in [1] by replacing the Euler-Maruyama scheme with the extended Milstein scheme (18), and this replacement does not require special treatments. We thus omit the detailed proof. □

We now return to the proof of Theorem 2. Since the upper bound for the term $\mathcal{E}^f_{4,n}(h,x)$ is provided in Lemma 4, we will study the other three terms. For the first term, we have $\mathcal{E}^f_{1,n}(h,x) = \mathcal{E}^{f,(\mathrm{I})}_{1,n}(h,x) + \mathcal{E}^{f,(\mathrm{II})}_{1,n}(h,x)$, with

$$\mathcal{E}^{f,(\mathrm{I})}_{1,n}(h,x) \equiv -h^2 \mathbb{E}\big[\Phi_3((n-1)h, X^x_{(n-1)h})\big],$$

$$\mathcal{E}^{f,(\mathrm{II})}_{1,n}(h,x) \equiv h^2 \sum_{k=0}^{n-1} \mathbb{E}\big[\Phi_3(kh, X^x_{kh})\big] - h \int_0^T \mathbb{E}\big[\Phi_3(s, X^x_s)\big]ds.$$

From the definition of Φ_3 and the bound (39) in Lemma 2, we have:

$$\big|\mathcal{E}^{f,(\mathrm{I})}_{1,n}(h,x)\big| \le K(T)\frac{\|f\|_\infty}{n^2}(1+|x|^Q). \tag{43}$$

We also have:

$$\Big|\frac{T}{n}\sum_{k=0}^{n-1} \mathbb{E}\big[\Phi_1(kT/n, X^x_{kT/n})\big] - \int_0^T \mathbb{E}\big[\Phi_1(s, X^x_s)\big]ds\Big| \le K(T)\frac{\|f\|_\infty}{n}(1+|x|^Q),$$

where we applied Itô formula as

$$\begin{aligned}\int_{kh}^{(k+1)h} \mathbb{E}\big[\Phi_3(s, X^x_s)\big]ds &= h\mathbb{E}\big[\Phi_3(kh, X^x_{kh})\big] \\ &\quad + \int_{kh}^{(k+1)h}\int_{kh}^{s} \mathbb{E}\big[\partial_t \Phi_3(s_1, X^x_{s_1})\big]ds_1 ds \\ &\quad + \int_{kh}^{(k+1)h}\int_{kh}^{s} \mathbb{E}\big[\mathcal{L}\Phi_3(s_1, X^x_{s_1})\big]ds_1 ds, \\ &\quad k = 0,\dots,n-1,\end{aligned}$$

and the last two terms in the right-hand side are bounded by $K(T)\frac{\|f\|_\infty}{n^2}(1+|x|^Q)$ due to (39). We thus obtain the upper bound $\big|\mathcal{E}^f_{1,n}(h,x)\big| \le K(T)\frac{\|f\|_\infty}{n^2}(1+|x|^Q)$.

Next, we study the second error term $\mathcal{E}^f_{2,n}(h,x)$. The term writes $\mathcal{E}^f_{2,n}(h,x) = \mathcal{E}^{f,(\mathrm{I})}_{2,n}(h,x) + \mathcal{E}^{f,(\mathrm{II})}_{2,n}(h,x)$ with

$$\mathcal{E}^{f,(\mathrm{I})}_{2,n}(h,x) \equiv h^2 \sum_{0\le k\le [n/2]} \Big\{\mathbb{E}\big[\Phi_3(kh, \bar{X}^{n,x}_{kh})\big] - \mathbb{E}\big[\Phi_3(kh, X^x_{kh})\big]\Big\};$$

$$\mathcal{E}^{f,(\mathrm{II})}_{2,n}(h,x) \equiv h^2 \sum_{[n/2]+1\le k\le n-2} \Big\{\mathbb{E}\big[\Phi_3(kh, \bar{X}^{n,x}_{kh})\big] - \mathbb{E}\big[\Phi_3(kh, X^x_{kh})\big]\Big\}.$$

We consider the term $\mathcal{E}_{2,n}^{f,(\mathrm{I})}(h,x)$. First, applying (40) and Lemma 4 to (37), we have that: for any $\varphi \in \mathscr{B}_b(\mathbb{R}^N)$, there exist a non-decreasing function $K(\cdot)$ and constants $q, Q > 0$ such that for all $s \in [0,T]$,

$$\left|\mathbb{E}[\varphi(X_s^x)] - \mathbb{E}[\varphi(\bar{X}_s^{n,x})]\right| \leq \tfrac{K(T)}{T^q} \tfrac{\|\varphi\|_\infty}{n}(1+|x|^Q). \tag{44}$$

Using the bound (44) with $\varphi = \Phi_3(kh,\cdot)$, we get

$$\begin{aligned}\left|\mathcal{E}_{2,n}^{f,(\mathrm{I})}(h,x)\right| &\leq h^2 \sum_{0\leq k \leq [\frac{n}{2}]} \tfrac{K_1(T)}{T^{q_1}} \cdot \tfrac{\|\Phi_3(kh,\cdot)\|_\infty}{n}(1+|x|^{Q_1}) \\ &\leq \tfrac{K_2(T)}{T^{q_2}} \cdot \tfrac{\|f\|_\infty}{n^2}(1+|x|^{Q_2}),\end{aligned} \tag{45}$$

for some non-decreasing functions K_1, K_2 and constants $q_1, q_2, Q_1, Q_2 > 0$ independent of n, T. We have also used the estimate (4.9) in [1] to bound $\|\Phi_3(kh,\cdot)\|_\infty$ in the last inequality. We next study the term $\mathcal{E}_{2,n}^{f,(\mathrm{II})}(h,x)$. We apply the error expression (37) to the term $\mathbb{E}\big[\Phi_3(kh,\bar{X}_{kh}^{n,x})\big] - \mathbb{E}\big[\Phi_3(kh,X_{kh}^x)\big]$ by replacing the test function $f(\cdot)$ with $\Phi_3(kh,\cdot)$, and then the term is ultimately bounded by $\frac{K(T)}{T^q}\frac{\|f\|_\infty}{n}(1+|x|^Q)$ due to Lemma 3 with some non-decreasing function $K(\cdot)$ and constants $q, Q > 0$ independent of n, T. This immediately leads to that the term $\big|\mathcal{E}_{2,n}^{f,(\mathrm{II})}(h,x)\big|$ is bounded as the right-hand side of (45). Thus, we have

$$\begin{aligned}\left|\mathcal{E}_{2,n}^{f}(h,x)\right| &\leq h^2 \sum_{k=0}^{n-2} \left|\mathbb{E}\big[\Phi_3(kh,\bar{X}_{kh}^{n,x})\big] - \mathbb{E}\big[\Phi_3(kh,X_{kh}^x)\big]\right| \\ &\leq K(T)\frac{\|f\|_\infty}{n^2}(1+|x|^Q).\end{aligned}$$

Finally, we study the term $\mathcal{E}_{3,n}^{f}(h,x)$. Notice that each term in $\widetilde{\Phi}_k^{(n)}(x)$ contains triple time integrals inside the expectation. Thus, Lemma 2 yields

$$\left|\mathcal{E}_{3,n}^{f}(h,x)\right| \leq \sum_{k=0}^{n-1} \left|\widetilde{\Phi}_k^{(n)}(x)\right| \leq K(T)\frac{\|f\|_\infty}{n^2}(1+|x|^Q), \tag{46}$$

and we conclude.

6 Conclusions

We have proposed a straightforward and effective first order weak scheme (18) for diffusion processes and compared its efficacy with other popular first order schemes, specifically, Euler-Maruyama/Misltein/truncated Milstein schemes. The proposed scheme is always explicit and can be simulated with Gaussian random variables, with the same number of random variables required by the Euler-Maruyama scheme or the truncated Milstein scheme. We have shown that the proposed scheme achieves the first order weak convergence, but its leading order error term involves fewer terms than other first order schemes. In particular, when the diffusion coefficients contain a small parameter ε, the discretization bias can be significantly reduced with the aid of ε, though the effect does not necessarily appear for EM and (truncated) Milstein schemes. We then carried out numerical experiments of Asian option pricing, which shows that the new scheme provides much superior accuracy to the other first order schemes and behaves nearly as the second order scheme under the setting of small diffusions. Notably, we have also observed that there are no significant differences in the performance between the EM and Milstein schemes in the experiment.

The application of the proposed time-discretization scheme is not limited to Monte-Carlo estimate for the expectation of the law of diffusions. Indeed, developing simple and effective discretization is useful in a much wider context that requires an approximate sampling of diffusion processes, e.g., parameter estimation of SDEs or filtering of diffusion processes when only partial coordinates are observed. For instance, [8] recently emphasized that the use of time-discretization with accurate weak approximation leads to efficient parameter estimation of diffusions when the so-called *Data augmentation* approach [16] is required to conduct Bayesian inference from low frequency observations. Due to the simple definition and its effective weak approximation, our proposed scheme can be incorporated into many computational/statistical methodologies (e.g. [3, 9]).

Acknowledgments Yuga Iguchi is supported by the Additional Funding Programme for Mathematical Sciences, delivered by EPSRC (EP/V521917/1) and the Heilbronn Institute for Mathematical Research. Toshihiro Yamada was supported by JST PRESTO (JPMJPR2029), Japan. We thank Prof. Shigeo Kusuoka (University of Tokyo) for grateful advice for the method of the paper.

Appendix 1: Proof of Lemma 1

Making use of the generator $\mathcal{L}$ given in (3) and $\partial_t u = -\mathcal{L}u$, we have from (10) that

$$\Phi^{\mathrm{EM}}(t,x) = -\frac{1}{2}\sum_{i,j=1}^{N} b^i(x)b^j(x)\partial_{ij}u(t,x) - \frac{1}{2}\sum_{i,j,k=1}^{N} b^i(x)a^{jk}(x)\partial_{ijk}u(t,x)$$
$$-\frac{1}{8}\sum_{i,j,k,l=1}^{N} a^{ij}(x)a^{kl}(x)\partial_{ijkl}u(t,x) + \frac{1}{2}\mathcal{L}^2u(t,x). \tag{47}$$

Since

$$\mathcal{L}^2 u(t,x) = \sum_{i,j=1}^{N} b^i(x)\big\{\partial_i b^j(x)\partial_j u(t,x) + b^j(x)\partial_{ij}u(t,x)\big\}$$

$$+ \frac{1}{2}\sum_{i,j,k=1}^{N} b^i(x)\big\{\partial_i a^{jk}(x)\partial_{jk}u(t,x) + a^{jk}(x)\partial_{ijk}u(t,x)\big\}$$

$$+ \frac{1}{2}\sum_{i,j,k=1}^{N} a^{ij}(x)\big\{\partial_{ij} b^k(x)\partial_k u(t,x) + \partial_i b^k(x)\partial_{jk}u(t,x)$$

$$+ \partial_j b^k(x)\partial_{ik}u(t,x) + b^k(x)\partial_{ijk}u(t,x)\big\}$$

$$+ \frac{1}{4}\sum_{i,j,k,l=1}^{N} a^{ij}(x)\big\{\partial_{ij} a^{kl}(x)\partial_{kl}u(t,x) + \partial_i a^{kl}(x)\partial_{jkl}u(t,x)$$

$$+ \partial_j a^{kl}(x)\partial_{ikl}u(t,x) + a^{kl}(x)\partial_{ijkl}u(t,x)\big\},$$

it follows that

$$\Phi^{\mathrm{EM}}(t,x) = \frac{1}{2}\sum_{i,j=1}^{N} b^i(x)\partial_i b^j(x)\partial_j u(t,x) + \frac{1}{4}\sum_{i,j,k=1}^{N} b^i(x)\partial_i a^{jk}(x)\partial_{jk}u(t,x)$$

$$+ \frac{1}{4}\sum_{i,j,k=1}^{N} a^{ij}(x)\partial_{ij} b^k(x)\partial_k u(t,x)$$

$$+ \frac{1}{2}\sum_{i,j,k=1}^{N} a^{ij}(x)\partial_i b^k(x)\partial_{jk}u(t,x)$$

$$+ \frac{1}{8}\sum_{i,j,k,l=1}^{N} a^{ij}(x)\partial_{ij} a^{kl}(x)\partial_{kl}u(t,x)$$

$$+ \frac{1}{4}\sum_{i,j,k,l=1}^{N} a^{ij}(x)\partial_i a^{kl}(x)\partial_{jkl}u(t,x). \tag{48}$$

Substituting

$$\partial_i a^{jk}(x) = \sum_{m=1}^{d}\Big\{\partial_i\sigma_m^j(x)\,\sigma_m^k(x) + \sigma_m^j(x)\partial_i\sigma_m^k(x)\Big\},$$

$$\partial_{ij} a^{kl}(x) = \sum_{m=1}^{d}\Big\{\partial_i\sigma_m^k(x)\,\partial_j\sigma_m^l(x) + \partial_j\sigma_m^k(x)\,\partial_i\sigma_m^l(x)$$

$$+ \partial_{ij}\sigma_m^k(x)\,\sigma_m^l(x) + \sigma_m^k(x)\,\partial_{ij}\sigma_m^l(x)\Big\}$$

into the right hand side of the Eq. (48), we have:

$$
\begin{aligned}
\Phi^{\mathrm{EM}}(t,x) &= \tfrac{1}{2}\sum_{i=1}^{N} L_0 b^i(x)\partial_i u(t,x) + \tfrac{1}{2}\sum_{i,j=1}^{N}\sum_{m=1}^{d}\sigma_m^i(x)L_m b^j(x)\partial_{ij}u(t,x)\\
&\quad + \tfrac{1}{2}\sum_{i,j=1}^{N}\sum_{m=1}^{d}\sigma_m^i(x)L_0\sigma_m^j(x)\partial_{ij}u(t,x)\\
&\quad + \tfrac{1}{4}\sum_{i,j=1}^{N}\sum_{m_1,m_2}^{d} L_{m_1}\sigma_{m_2}^i(x)L_{m_1}\sigma_{m_2}^j(x)\partial_{ij}u(t,x)\\
&\quad + \tfrac{1}{2}\sum_{i,j,k=1}^{N}\sum_{m_1,m_2}^{d} L_{m_1}\sigma_{m_2}^i(x)\sigma_{m_1}^j(x)\sigma_{m_2}^k(x)\partial_{ijk}u(t,x)\\
&= \Phi_1(t,x)+\Phi_2(t,x)+\Phi_3(t,x),
\end{aligned}
$$

where the last line is deduced from:

$$
\begin{aligned}
&\tfrac{1}{4}\sum_{i,j=1}^{N}\sum_{m_1,m_2}^{d} L_{m_1}\sigma_{m_2}^i(x)L_{m_1}\sigma_{m_2}^j(x)\partial_{ij}u(t,x)\\
&\quad= \tfrac{1}{8}\sum_{i,j=1}^{N}\sum_{m_1,m_2}^{d} L_{m_1}\sigma_{m_2}^i(x)\Big\{L_{m_1}\sigma_{m_2}^j(x)+L_{m_2}\sigma_{m_1}^j(x)\Big\}\partial_{ij}u(t,x)\\
&\quad+ \tfrac{1}{8}\sum_{i,j=1}^{N}\sum_{m_1,m_2}^{d} L_{m_1}\sigma_{m_2}^i(x)\Big\{L_{m_1}\sigma_{m_2}^j(x)-L_{m_2}\sigma_{m_1}^j(x)\Big\}\partial_{ij}u(t,x),
\end{aligned}
$$

and we conclude.

Appendix 2: Proof of Proposition 2

It holds that for $k=0,1,\ldots,n-2$,

$$
\begin{aligned}
&(Q_h)^k(P_h-Q_h)P_{T-(k+1)h}f(x)\\
&\quad= \mathbb{E}\big[P_{T-kh}f(\bar{X}_{kh}^{n,x})\big]-\mathbb{E}\Big[P_{T-(k+1)h}f\big(\bar{X}_{(k+1)h}^{n,x}\big)\Big] = -\mathbb{E}\Big[F_k^{(n)}(\bar{X}_{kh}^{n,x})\Big],
\end{aligned} \tag{49}
$$

where we have set:

$$F_k^{(n)}(z) \equiv \mathbb{E}\Big[u((k+1)h, \bar{X}_h^z) - u(kh, z)\Big], \quad z \in \mathbb{R}^N,$$

with $\bar{X}_h^z$ being the one step extended Milstein scheme (33) given the previous state z. We now consider the expansion of $F_k^{(n),x}(z)$ via iterative application of Itô's formula to $u((k+1)h, \bar{X}_h^z)$. Application of Itô's formula yields:

$$F_k^{(n)}(z) = T_1 + T_2 + T_3,$$

where we have defined:

$$T_1 = \mathbb{E}\Big[\int_0^h \partial_t u(kh+s, \bar{X}_s^z)ds\Big], \qquad T_2 = \sum_{i=1}^N \mathbb{E}\Big[\int_0^h \partial_i u(kh+s, \bar{X}_s^z)d\bar{X}_s^{z,i}\Big],$$

$$T_3 = \tfrac{1}{2}\sum_{i,j=1}^N \mathbb{E}\Big[\int_0^h \partial_{ij} u(kh+s, \bar{X}_s^z)d\langle \bar{X}^{z,i}, \bar{X}^{z,j}\rangle_s\Big].$$

We have that for $t > 0$ and $x \in \mathbb{R}^N$,

$$\begin{aligned} d\bar{X}_t^{x,i} &= \sum_{k=0}^d \sigma_k^i(x)dB_t^k \\ &+ \tfrac{1}{2}\sum_{0\le k_1,k_2\le d} \big\{L_{k_1}\sigma_{k_2}^i(x) + L_{k_2}\sigma_{k_1}^i(x)\big\}B_t^{k_1}dB_t^{k_2}, \ 1 \le i \le N, \end{aligned}$$

$$\begin{aligned} d\langle \bar{X}_\cdot^{x,i}, \bar{X}_\cdot^{x,j}\rangle_t &= a^{ij}(x)dt + \tfrac{1}{2}\sum_{\substack{1\le k_1\le d\\ 0\le k_2\le d}} \sigma_{k_1}^i(x)\{L_{k_1}\sigma_{k_2}^j(x) + L_{k_2}\sigma_{k_1}^j(x)\}B_t^{k_2}dt \\ &+ \tfrac{1}{2}\sum_{\substack{1\le k_1\le d\\ 0\le k_2\le d}} \sigma_{k_1}^j(z)\{L_{k_1}\sigma_{k_2}^i(x) + L_{k_2}\sigma_{k_1}^i(x)\}B_t^{k_2}dt \\ &+ \tfrac{1}{4}\sum_{\substack{0\le k_1,k_2\le d\\ 1\le k_3\le d}} \big\{L_{k_1}\sigma_{k_3}^i(x) + L_{k_3}\sigma_{k_1}^i(x)\big\} \\ &\big\{L_{k_2}\sigma_{k_3}^j(x) + L_{k_3}\sigma_{k_2}^j(x)\big\}B_t^{k_1}B_t^{k_2}dt, \quad 1 \le i, j \le N. \end{aligned}$$

For the term T_1, the stochastic Taylor expansion gives:

$$
\begin{aligned}
T_1 &= h \cdot \partial_t u(kh, z) + \mathbb{E}\Big[\int_0^h \int_0^s \partial_t^2 u(kh+v, \bar{X}_v^z) dv ds\Big] \\
&\quad + \sum_{i=1}^N \mathbb{E}\Big[\int_0^h \int_0^s \partial_i \partial_t u(kh+v, \bar{X}_v^z) d\bar{X}_v^{z,i} ds\Big] \\
&\quad + \tfrac{1}{2} \sum_{i,j=1}^N \mathbb{E}\Big[\int_0^h \int_0^s \partial_{ij} \partial_t u(kh+v, \bar{X}_v^z) d\langle \bar{X}^{z,i}, \bar{X}^{z,j}\rangle_v ds\Big] \\
&= h \cdot \partial_t u(kh, z) + \tfrac{h^2}{2} \cdot \partial_t^2 u(kh, z) + \tfrac{h^2}{2} \cdot L_0 \partial_t u(kh, z) + \mathscr{R}_{k,1}^{(n)}(h, z),
\end{aligned}
$$

where $\mathscr{R}_{k,1}^{(n)}(h, z)$ is the remainder term such that $\mathbb{E}[\mathscr{R}_{k,1}^{(n)}(h, z)]$ is given in the form of (36).

For the second term T_2, we have:

$$
\begin{aligned}
T_2 &= \sum_{i=1}^N \mathbb{E}\Big[\int_0^h \partial_i u(kh+s, \bar{X}_s^z) b^i(z) ds\Big] \\
&\quad + \tfrac{1}{2} \sum_{i=1}^N \sum_{k=0}^d \mathbb{E}\Big[\int_0^h \partial_i u(kh+s, \bar{X}_s^z)\big\{L_k \sigma_0^i(z) + L_0 \sigma_k^i(z)\big\} B_s^k ds\Big] \\
&= \sum_{i=1}^N \Big\{h \cdot \partial_i u(kh, z) + \tfrac{h^2}{2} \cdot \partial_t \partial_i u(kh, z) + \tfrac{h^2}{2} \cdot L_0 \partial_i u(kh, z)\Big\} b^i(z) \\
&\quad + \tfrac{h^2}{2} \sum_{i=1}^N \partial_i u(kh, z) L_0 \sigma_0^i(z) \\
&\quad + \tfrac{h^2}{4} \sum_{i,j=1}^N \sum_{k_1=1}^d \partial_{ij} u(kh, z) \sigma_{k_1}^i(z) \big\{L_{k_1} \sigma_0^j(z) + L_0 \sigma_{k_1}^j(z)\big\} + \mathscr{R}_{k,2}^{(n)}(h, z),
\end{aligned}
$$

where $\mathscr{R}_{k,2}^{(n)}(h, z)$ is the remainder term such that $\mathbb{E}[\mathscr{R}_{k,2}^{(n)}(h, z)]$ is given in the form of (36).

Finally, for the term T_3 we have:

$$
\begin{aligned}
T_3 &= \tfrac{1}{2} \sum_{i,j=1}^N \mathbb{E}\Big[\int_0^h \partial_{ij} u(kh+s, \bar{X}_s^z) a^{ij}(z) ds\Big] \\
&\quad + \tfrac{1}{2} \sum_{i,j=1}^N \sum_{\substack{1 \le k_1 \le d \\ 0 \le k_2 \le d}} \mathbb{E}\Big[\int_0^h \partial_{ij} u(kh+s, \bar{X}_s^z) \sigma_{k_1}^i(z)
\end{aligned}
$$

$$\{L_{k_1}\sigma^j_{k_2}(z)+L_{k_2}\sigma^j_{k_1}(z)\}B^{k_2}_s ds\Big]$$

$$+\tfrac{1}{8}\sum_{i,j=1}^{N}\sum_{\substack{0\le k_1,k_2\le d\\1\le k_3\le d}}\mathbb{E}\Big[\int_0^h \partial_{ij}u(kh+s,\bar{X}^z_s)\{L_{k_1}\sigma^i_{k_3}(z)+L_{k_3}\sigma^i_{k_1}(z)\}$$

$$\{L_{k_2}\sigma^j_{k_3}(z)+L_{k_3}\sigma^j_{k_2}(z)\}B^{k_1}_s B^{k_2}_s ds\Big]$$

$$=\tfrac{h}{2}\sum_{i,j=1}^{N}\partial_{ij}u(kh,z)a^{ij}(z)+\tfrac{h^2}{4}\sum_{i,j=1}^{N}\partial_t\partial_{ij}u(kh,z)a^{ij}(z)$$

$$+\tfrac{h^2}{4}\sum_{i,j,l=1}^{N}\partial_{ijl}u(kh,z)a^{ij}(z)b^l(z)$$

$$+\tfrac{h^2}{8}\sum_{i,j,m,l=1}^{N}\partial_{ijml}u(kh,z)a^{ij}(z)a^{ml}(z)$$

$$+\tfrac{h^2}{4}\sum_{i,j=1}^{N}\sum_{1\le k_1\le d}\partial_{ij}u(kh,z)\sigma^i_{k_1}(z)\{L_{k_1}\sigma^j_0(z)+L_0\sigma^j_{k_1}(z)\}$$

$$+\tfrac{h^2}{2}\sum_{i,j,l=1}^{N}\sum_{1\le k_1,k_2\le d}\partial_{ijl}u(kh,z)\sigma^i_{k_1}(z)\sigma^l_{k_2}(z)L_{k_1}\sigma^j_{k_2}(z)$$

$$+\tfrac{h^2}{8}\sum_{i,j=1}^{N}\sum_{k_1,k_2=1}^{d}\partial_{ij}u(kh,z)\{L_{k_1}\sigma^i_{k_2}(x)L_{k_1}\sigma^j_{k_2}(x)$$

$$+L_{k_1}\sigma^i_{k_2}(x)L_{k_2}\sigma^j_{k_1}(x)\}+\mathscr{R}^{(n)}_{k,3}(h,z),$$

where $\mathscr{R}^{(n)}_{k,3}(h,z)$ is such that $\mathbb{E}[\mathscr{R}^{(n)}_{k,3}(h,z)]$ is given in the form of (36). Here, we introduce the differential operator: for $\varphi\in C^\infty_b(\mathbb{R}^N;\mathbb{R}))$, $\xi, x\in\mathbb{R}^N$,

$$L^\xi_0\varphi(x)=\sum_{i=1}^{N}b^i(\xi)\partial_i\varphi(x)+\tfrac{1}{2}\sum_{i,j=1}^{N}a^{ij}(\xi)\partial_{ij}\varphi(x). \tag{50}$$

Making use of $\partial_t u = L_0 u$ and the above operator L^ξ_0, we obtain:

$$F^{(n)}_k(z)=h\cdot\partial_t u(kh,z)+h\cdot L_0u(kh,z)+\tfrac{h^2}{2}L_0\partial_t u(kh,z)$$

$$+\tfrac{h^2}{2}L_0L^\xi_0u(kh,z)|_{\xi=z}+\tfrac{h^2}{2}\sum_{i=1}^{N}\partial_i u(kh,z)L_0\sigma^i_0(z)$$

$$
\begin{aligned}
&+ \tfrac{h^2}{2} \sum_{i,j=1}^{N} \sum_{k_1=1}^{d} \partial_{ij} u(kh, z) \sigma^i_{k_1}(z) \big\{ L_{k_1} \sigma^j_0(z) + L_0 \sigma^j_{k_1}(z) \big\} \\
&+ \Phi_2(kh, z) + \mathscr{R}^{(n)}_k(h, z) \\
&= -\tfrac{h^2}{2} L_0^2 u(kh, z) + \tfrac{h^2}{2} L_0 L_0^{\xi} u(kh, z)|_{\xi = z} + \Phi_1(kh, z) \\
&+ \Phi_2(kh, z) + \mathscr{R}^{(n)}_k(h, z),
\end{aligned}
$$

where we have defined $\mathscr{R}^{(n)}_k(h, z) = \sum_{i=1}^{3} \mathscr{R}^{(n)}_{k,i}(h, z)$. Since it follows from (47) that

$$
\Phi^{\mathrm{EM}}(kh, z) = \tfrac{h^2}{2} L_0^2 u(kh, z) - \tfrac{h^2}{2} L_0 L_0^{\xi} u(kh, z)|_{\xi = z}, \tag{51}
$$

we have

$$
\begin{aligned}
F^{(n)}_k(z) &= -\Phi^{\mathrm{EM}}(kh, z) + \Phi_1(kh, z) + \Phi_2(kh, z) + \mathscr{R}^{(n)}_k(h, z) \\
&= -\Phi_3(kh, z) + \mathscr{R}^{(n)}_k(h, z).
\end{aligned} \tag{52}
$$

From (49) and (52), we conclude.

Appendix 3: Proof of Proposition 1

Before showing Proposition 1, we prepare some tools from Malliavin calculus. The details are found in, e.g. [15].

Let $(\Omega, \mathcal{F}, \mu)$ be d-dimensional Wiener space and $H = L^2([0, T]; \mathbb{R}^d)$ equipped with the inner product $\langle \cdot, \cdot \rangle_H$ defined as $\langle h_1, h_2 \rangle_H = \int_0^T h_1(s) \cdot h_2(s) ds$, $h_1, h_2 \in H$. For $h \in H$, we write $W(h) \equiv \int_0^T h(s) \cdot dW_s$ and introduce a space of some Wiener functionals as $\mathscr{S} \equiv \big\{ F = f\big(W(h_1), \ldots, W(h_n)\big) \mid f \in C_p^{\infty}(\mathbb{R}^n; \mathbb{R}),\ h_1, \ldots, h_n \in H,\ n \geq 1 \big\}$. For $p \geq 1$, the Malliavin derivative operator $D : \mathscr{S} \to L^p(\Omega; H)$ is defined via the following operation:

$$
\mathscr{S} \ni F \mapsto\ D_t F = \sum_{i=1}^{n} \partial_i f(W(h_1), \ldots, W(h_n)) h_i(t),\ t \in [0, T].
$$

For $F \in \mathscr{S}$, $\{D_t F\}_{t \geq 0}$ is typically treated as a d-dimensional stochastic process, and we write $D_{j,t} F$, $1 \leq j \leq d$ as the j-th element of $D_t F$. For $k \in \mathbb{N}$, the k-th order Malliavin derivative is denoted by D^k which is closable from $\mathscr{S}$ into $L^p(\Omega; H^{\otimes k})$. We write $\mathbb{D}_{k,p}$, $k \in \mathbb{N}$, $p \geq 1$ as the completion of $\mathscr{S}$ w.r.t. the norm

$$
\|F\|_{k,p} = \big\{ \mathbb{E}[|F|^p] + \sum_{j=1}^{k} \mathbb{E}[\|D^j F\|^p_{H^{\otimes k}}] \big\}^{1/p},\ \ F \in \mathscr{S}.
$$

Define $\mathbb{D}^{\infty} = \bigcap_{k \in \mathbb{N}} \bigcap_{p \geq 1} \mathbb{D}_{k,p}$. For a m-dimensional random vector $F = (F^1, \ldots, F^m) \in (\mathbb{D}^{\infty})^m$, the Malliavin covariance $\sigma^F = (\sigma^F_{ij})_{1 \leq i,j \leq m}$ is defined as $\sigma^F_{ij} = \langle DF^i, DF^j \rangle_H$. Then, F is said to be *non-degenerate in Malliavin sense* if σ^F is invertible a.s. and also $(\det \sigma^F)^{-1} \in \bigcap_{p \geq 1} L^p(\Omega)$. For non-degenerate Wiener functionals, we have the integration by parts on Wiener space as follows: let $G \in \mathbb{D}^{\infty}$, $\varphi \in C_b^{\infty}(\mathbb{R}^m)$ and $F \in (\mathbb{D}^{\infty})^m$ be non-degenerate in Malliavin sense. Then, for any multi-index $\alpha \in \{1, \ldots, m\}^k$, $k \in \mathbb{N}$, there exists $\mathscr{H}_{\alpha}(F, G)$ such that

$$\mathbb{E}[\partial_{\alpha} \varphi(F) G] = \mathbb{E}[\varphi(F) \mathscr{H}_{\alpha}(F, G)] \tag{53}$$

where the stochastic weight $\mathscr{H}_{\alpha}(F, G)$ is recursively defined via:

$$\mathscr{H}_{(\alpha_1)}(F, G) = \sum_{j=1}^{m} \delta\big((\sigma^F)^{-1}_{\alpha_1 j} G\, DF\big), \quad \mathscr{H}_{(\alpha_1, \ldots, \alpha_k)}(F, G) = \mathscr{H}_{(\alpha_k)}\big(F, \mathscr{H}_{(\alpha_1, \ldots, \alpha_{k-1})}(F, G)\big)$$

with δ denoting the adjoint operator of D.

Proof of Proposition 1 We recall $u^{\varepsilon}(s, x) = \mathbb{E}\big[f(X^{\varepsilon,x}_{T-s})\big]$ and set $\alpha \in \{1, \ldots, N\}^k$, $k \in \mathbb{N}$ throughout the proof. We have that

$$\Big| \int_0^T \mathbb{E}\big[g(X^{x,\varepsilon}_s) \partial_{\alpha} u^{\varepsilon}(s, X^{x,\varepsilon}_s)\big] ds \Big| = J_1 + J_2, \tag{54}$$

where

$$J_1 = \Big| \int_0^{T/2} \mathbb{E}\big[g(X^{x,\varepsilon}_s) \partial_{\alpha} u^{\varepsilon}(s, X^{x,\varepsilon}_s)\big] ds \Big|,$$
$$J_2 = \Big| \int_{T/2}^{T} \mathbb{E}\big[g(X^{x,\varepsilon}_s) \partial_{\alpha} u^{\varepsilon}(s, X^{x,\varepsilon}_s)\big] ds \Big|. \tag{55}$$

We then derive upper bounds for the terms J_1 and J_2. We first consider the term J_1. We note that $X^{\xi,\varepsilon}_{T-s}$ is non-degenerate in Malliavin sense under the condition (H1) and (H2). Thus, applying the Malliavin integration by parts (53), we have that: for any $(s, \xi) \in [0, T] \times \mathbb{R}^N$ and $1 \leq i \leq N$,

$$\partial_{\xi_i} u^{\varepsilon}(s, \xi) = \sum_{j=1}^{N} \mathbb{E}\big[\partial_j f(X^{\xi,\varepsilon}_{T-s}) J^{\xi,\varepsilon}_{T-s,ij}\big] = \sum_{j=1}^{N} \mathbb{E}\big[f(X^{\xi,\varepsilon}_{T-s}) \mathscr{H}_{(j)}\big(X^{\xi,\varepsilon}_{T-s}, J^{\xi,\varepsilon}_{T-s,ij}\big)\big], \tag{56}$$

where $J^{\xi,\varepsilon}_{T-s} = \left(J^{\xi,\varepsilon}_{T-s,ij}\right)_{1\le i,j\le N}$ is defined as $J^{\xi,\varepsilon}_{T-s,ij} = \frac{\partial X^{\xi,\varepsilon,j}_{T-s}}{\partial \xi_i}$. We have that $J^{\xi,\varepsilon}_{T-s,ij} \in \mathbb{D}^\infty$ and $\|J^{\xi,\varepsilon}_{T-s,ij}\|_{k,p} < \infty$ for any $k \in \mathbb{N}$ and $p \ge 1$ due to the condition (H1). Then, it holds that

$$\left|\partial_\alpha u^\varepsilon(s,\xi)\right| \le \frac{K(T)}{(T-s)^q\varepsilon^k}\|f\|_\infty(1+|\xi|^Q) \tag{57}$$

for some constants $q, Q > 0$, $p \ge 1$ and non-decreasing function $K(\cdot)$ independent of x and ε. To obtain the bound (57), we have exploited the following estimate whose proof is postponed to the end of this section:

Lemma 5 *Let $x \in \mathbb{R}^N$ and $G \in \mathbb{D}^\infty$ satisfying that: for any $j \in \mathbb{N}$ and $q \ge 1$, there exists a constant such that $\|G\|_{j,p} \le c$ for some $c > 0$. Assume the conditions* (H1) *and* (H2) *with $M = 1$ hold. Then, for any $\beta \in \{1,\dots,N\}^k$, $k \in \mathbb{N}$ and $p \ge 1$, there exist constants $q, Q > 0$ and a non-decreasing function $K(\cdot)$ independent of x and $\varepsilon \in (0,1)$ such that*

$$\left\{\mathbb{E}\left[\left|\mathscr{H}_{(\beta)}(X^{x,\varepsilon}_t, G)\right|^p\right]\right\}^{1/p} \le \frac{K(t)}{t^q\varepsilon^k}(1+|x|^Q). \tag{58}$$

Using the bound (57) and the polynomial growth of g with moment bound for X^x_s under (H1), we have that:

$$J_1 \le \frac{K(T)}{T^q\varepsilon^k}\|f\|_\infty \times \sup_{s\in[0,T]} \mathbb{E}[(1+|X^{x,\varepsilon}_{T-s}|^{Q_1})] \le \frac{K(T)}{T^q\varepsilon^k}\|f\|_\infty(1+|x|^{Q_2}) \tag{59}$$

for some constants $C_1, q, Q_1, Q_2 > 0$ and non-decreasing functions $K(\cdot)$ which are independent of x and ε. For the second term J_2, we apply the Malliavin integration by parts and Lemma 5 to obtain that:

$$\begin{aligned} J_2 &= \left|\int_{T/2}^T \mathbb{E}\left[u^\varepsilon(s, X^{x,\varepsilon}_s)\mathscr{H}_{(\alpha)}\left(X^{x,\varepsilon}_s, g(X^{x,\varepsilon}_s)\right)\right]ds\right| \\ &\le \|f\|_\infty \int_{T/2}^T \mathbb{E}\left[\left|\mathscr{H}_{(\alpha)}\left(X^{x,\varepsilon}_s, g(X^{x,\varepsilon}_s)\right)\right|\right]ds \le \frac{K(T)}{T^q\varepsilon^k}\|f\|_\infty(1+|x|^Q), \end{aligned} \tag{60}$$

for some constants $q, Q > 0$ and a non-decreasing function K independent of $x \in \mathbb{R}^N$ and $\varepsilon \in (0,1)$, where in the last line we have used the polynomial growth of g and the following estimate under the condition (H1): for any $k \in \mathbb{N}$ and $p \ge 1$ there exist constants $c, Q > 0$ independent of s, x, ε such that $\|g(X^{x,\varepsilon}_s)\|_{k,p} < c(1+|x|^Q)$. The proof of Proposition 1 is now complete.

Proof of Lemma 5

We have from (2.32) in [15] that: for $1 \leq p \leq q < \infty$,

$$\big\{\mathbb{E}\big[|\mathscr{H}_{(\beta)}(X_t^{x,\varepsilon}, G)|^p\big]\big\}^{1/p} \leq C_1 \big\| \gamma^{X_t^{x,\varepsilon}} DX_t^{x,\varepsilon} \big\|_{k,\gamma}^k \|G\|_{k,q} \leq C_2 \big\| \gamma^{X_t^{x,\varepsilon}} DX_t^{x,\varepsilon} \big\|_{k,\gamma}^k, \tag{61}$$

with some constants $C_1, C_2 > 0$ and $\gamma > 1$. The Malliavin derivative of $X_t^{x,\varepsilon,i}$, $1 \leq i \leq N$, is given as (for instance see [15, 17]):

$$D_{j,s} X_t^{x,\varepsilon,i} = \varepsilon \big[J_t^{x,\varepsilon} \big(J_s^{x,\varepsilon}\big)^{-1} \sigma_j(X_s^{x,\varepsilon}) \big]_i, \qquad s \in [0,t], \;\; 1 \leq j \leq d. \tag{62}$$

We define $N \times N$ matrix valued random variable $\widetilde{M}^{X_t^{x,\varepsilon}} = \big(\widetilde{M}_{i_1 i_2}^{X_t^{x,\varepsilon}}\big)_{1 \leq i_1, i_2 \leq N}$ as:

$$\widetilde{M}^{X_t^{x,\varepsilon}} = \sum_{j=1}^{d} \int_0^t J_t^{x,\varepsilon} \big(J_s^{x,\varepsilon}\big)^{-1} V_j(X_s^{x,\varepsilon}) \otimes J_t^{x,\varepsilon} \big(J_s^{x,\varepsilon}\big)^{-1} V_j(X_s^{x,\varepsilon}) ds. \tag{63}$$

Under the uniform Hörmander condition (H2) with $M = 1$, the matrix $\widetilde{M}^{X_t^{x,\varepsilon}}$ is shown to be invertible a.s. and also $\mathbb{E}\big[\big|(\det \widetilde{M}^{X_t^{x,\varepsilon}})^{-1}\big|^p\big] < \infty$ for all $p \geq 1$. We then write the inverse of the matrix $\widetilde{M}^{X_t^{x,\varepsilon}}$ as $\widetilde{\gamma}^{X_t^{x,\varepsilon}}$. By considering stochastic Taylor expansion of $\big(J_s^{x,\varepsilon}\big)^{-1} V_j(X_s^{x,\varepsilon})$, it is shown from Theorem 6.16 (and its proof) in [17] that: there exist constants $q, Q > 0$ and a non-decreasing function $K(\cdot)$ independent of x, ε, t such that

$$\big\| \widetilde{\gamma}_{i_1 i_2}^{X_t^{x,\varepsilon}} \big\|_{m,n} \leq \frac{K(t)}{t^q} (1 + |x|^Q), \qquad m, n \in \mathbb{N}, \quad 1 \leq i_1, i_2 \leq N. \tag{64}$$

In particular, noticing that the Malliavin covariance of $X_t^{x,\varepsilon}$ is given as $M^{X_t^{x,\varepsilon}} = \varepsilon^2 \widetilde{M}^{X_t^{x,\varepsilon}}$, we have that: for $m, n \in \mathbb{N}$ and $1 \leq i_1, i_2 \leq N$,

$$\big\| \gamma_{i_1 i_2}^{X_t^{x,\varepsilon}} \big\|_{m,n} = \varepsilon^{-2} \big\| \widetilde{\gamma}_{i_1 i_2}^{X_t^{x,\varepsilon}} \big\|_{m,n} \leq \frac{K(t)}{t^q \varepsilon^2} (1 + |x|^Q). \tag{65}$$

We thus obtain from (62), (65) and Hölder's inequality that:

$$\big\| \gamma^{X_t^{x,\varepsilon}} DX_t^{x,\varepsilon} \big\|_{k,\gamma}^k \leq \frac{K(t)}{t^q \varepsilon^k} (1 + |x|^Q), \tag{66}$$

and we conclude.

References

1. Bally, V., Talay, D.: The law of the Euler scheme for stochastic differential equations. Probab. Theory Relat. Fields **104**(1), 43–60 (1996)
2. Giles, M.B., Szpruch, L.: Antithetic multilevel Monte Carlo estimation for multi-dimensional SDEs without Lévy area simulation. Ann. Appl. Probab. **24**(4), 1585–1620 (2014)
3. Graham, M.M., Thiery, A.H., Beskos, A.: Manifold Markov chain Monte Carlo methods for Bayesian inference in diffusion models. J. R. Stat. Soc. Ser. B Stat. Methodol. **84**(4), 1229–1256 (2022)
4. Gyurkó, L.G., Lyons, T.: Rough paths based numerical algorithms in computational finance. Contemp. Math. **515**, 17–46 (2010)
5. Gyurkó, L.G., Lyons, T.J.: Efficient and practical implementations of cubature on Wiener space. Stoch. Anal. **2010**, 73–111 (2011)
6. Iguchi, Y., Yamada, T.: Operator splitting around Euler–Maruyama scheme and high order discretization of heat kernels. Math. Modell. Numer. Anal. **55**, S323–S367 (2021)
7. Iguchi, Y., Yamada, T.: A second-order discretization for degenerate systems of stochastic differential equations. IMA J. Numer. Anal. **41**(4), 2782–2829 (2021)
8. Iguchi, Y., Beskos, A., Graham, M.M.: Parameter estimation with increased precision for elliptic and hypo-elliptic diffusions. Bernoulli. **31**(1), 333–358 (2025)
9. Iguchi, Y., Jasra, A., Maama, M., Beskos, A.: Antithetic multilevel methods for elliptic and hypo-elliptic diffusions with applications. SIAM-ASA J. Uncertain. Quantif. **13**(2), 805–830 (2025)
10. Kloeden, P.E., Platen, E.: Numerical Solution of Stochastic Differential Equations. Springer, Berlin (1992)
11. Litterer, C., Lyons, T.: High order recombination and an application to cubature on Wiener space. Ann. Appl. Probab. 22(4), 1301–1327 (2012)
12. Milstein, G.N., Tretyakov, M.V.: Stochastic Numerics for Mathematical Physics. Springer Cham (2021)
13. Naito, R., Yamada, T.: A third-order weak approximation of multidimensional Itô stochastic differential equations. Monte Carlo Methods Appl. **25**(2), 97–120 (2019)
14. Ninomiya, S., Victoir, N.: Weak approximation of stochastic differential equations and application to derivative pricing. Appl. Math. Finance **15**(2), 107–121 (2008)
15. Nualart, D.: The Malliavin Calculus and Related Topics, vol. 1995. Springer, Berlin (2006)
16. Papaspiliopoulos, O., Roberts, G.O., Stramer, O.: Data augmentation for diffusions. J. Comput. Graph. Stat. **22**(3), 665–688 (2013)
17. Shigekawa, I.: Stochastic Analysis, vol. 224. American Mathematical Society, Providence (2004)
18. Takahashi, A., Yamada, T.: An asymptotic expansion with push-down of Malliavin weights. SIAM J. Financ. Math. **3**, 95–136 (2012)
19. Takahashi, A., Yamada, T.: A weak approximation with asymptotic expansion and multidimensional Malliavin weights. Ann. Appl. Probab. **26**(2), 818–856 (2016)
20. Talay, D.: Efficient numerical schemes for the approximation of expectations of functionals of the solution of a SDE, and applications. In: *Filtering and Control of Random Processes: Proceedings of the ENST-CNET Colloquium Paris, France, February 23–24, 1983*, pages 294–313. Springer, Berlin (1984)
21. Yamada, T.: An arbitrary high order weak approximation of SDE and Malliavin Monte Carlo: analysis of probability distribution functions. SIAM J. Numer. Anal. **57**(2), 563–591 (2019)
22. Yamada, T.: Total variation bound for Milstein scheme without iterated integrals. Monte Carlo Methods Appl. **29**(3), 221–242 (2023)
23. Yamada, T., Yamamoto, K.: A second-order discretization with Malliavin weight and Quasi-Monte Carlo method for option pricing. Quant. Finance **20**(11), 1825–1837 (2020)

Expansion of Bermudan Option Price Using Deep Learning

Riu Naito and Toshihiro Yamada

Abstract The paper proposes a new pricing algorithm for high-dimensional Bermudan options based on an expansion method of Wiener functionals and deep learning which is a version of weak approximation scheme of *Naito and Yamada, Journal of Computational Finance* (2024). The proposed method gives efficient and accurate approximations for Bermudan option prices even in a high-dimensional setting. The theoretical rate of convergence of the proposed method is provided. Numerical experiments including a 100-dimensional stochastic volatility model confirm the validity of the proposed method.

Keywords Bermudan option · Deep learning · Asymptotic expansion

MSC Classification 60H07, 65C30, 68T07, 91G20, 91G60

1 Introduction

Early-exercisable options such as Bermudan options are widely used in financial industry since such products provide important hedging instruments in the business. While fast and efficient numerical scheme are necessary in Bermudan option pricing, the values of options under high-dimensional models are difficult to estimate because iterative functional approximations of nested conditional expectations are required.

R. Naito
University of Toyama, Toyama, Japan
e-mail: riunaito@eco.u-toyama.ac.jp

T. Yamada (✉)
Graduate School of Economics, Hitotsubashi University, Tokyo, Japan
e-mail: toshihiro.yamada@r.hit-u.ac.jp

J. Akahori et al. (eds.), *Recent Developments in Stochastic Numerics and Computational Finance*, ICIAM2023 Springer Series 6,
https://doi.org/10.1007/978-981-95-0652-1_3

In the literature on Bermudan option pricing, various numerical methods have been developed. One of the widely adopted approaches is the regression-based method introduced in [9, 17, 27], which are further investigated by Belomestny [6] and Clément et al. [7]. Another conventional method is the dual approach developed in [1, 11, 21]. In [2], the quantization method is introduced.

There has been recent development in deep learning-based algorithms aimed at solving nonlinear problems and pricing early-exercisable financial derivatives; see [3–5, 8, 10, 12, 16, 19, 22, 26]. In [4], they propose new algorithms, called the deep optimal stopping, which approximate the optimal stopping time directly by using deep learning. The deep learning-based regression method is proposed in [16]. In [8], the authors introduce a deep learning-based algorithm for pricing early-exercisable options by extending the algorithms of [10].

To obtain fast and accurate pricing scheme for Bermudan options, it is important to reduce the approximation error for the continuation value at each early-exercise date. With the aim of obtaining accurate approximations, in [19], the authors propose an arbitrary order discretization (weak approximation) scheme in Bermudan option pricing, based on [13, 18, 29].

In this paper, we introduce a new efficient deep learning-based algorithm for pricing Bermudan options based on an expansion method of Wiener functionals [23–25, 28]. We also provide the theoretical convergence rate and check it through numerical experiments including 100-dimensional stochastic volatility model. The deep learning-based algorithm can be efficiently implemented for high-dimensional problems which is difficult to compute by the classical algorithms such as [17] due to the computational cost. In this sense, the proposed algorithm is an alternative approach for high-dimensional Bermudan option pricing. Moreover, we note that the proposed method has another advantage compared to [19], that is, it can be implementable without any computational cost due to discretization of the intervals of the early-exercise dates.

The organization of the paper is as follows. In Sect. 2, we introduce the computational settings in the paper. The main result and the proposed algorithms are provided in Sect. 3. Section 4 shows numerical examples including a 100-dimensional stochastic volatility model to validate the efficiency of the algorithm. We conclude in Sect. 5. Appendix are devoted to the mathematical proofs of our main result.

2 Setting

We first introduce notations. Let $C_b^\infty(\mathbb{R}^m; \mathbb{R}^n)$ be the set of all infinitely continuously differentiable functions $f : \mathbb{R}^m \to \mathbb{R}^n$ such that f and its partial derivatives of all order are bounded. We write $C_b^\infty(\mathbb{R}^m)$ for $C_b^\infty(\mathbb{R}^m; \mathbb{R})$. Analogously, let $C_b^\infty([0, T]\times\mathbb{R}^m; \mathbb{R}^n)$ be the set of all infinitely continuously differentiable functions $f : [0, T] \times \mathbb{R}^m \to \mathbb{R}^n$ such that f and its partial derivatives of all order are bounded. We write $C_b^\infty(\mathbb{R}^m)$ for $C_b^\infty(\mathbb{R}^m;\mathbb{R})$. For a bounded measurable function $f:\mathbb{R}^m\to\mathbb{R}$, we define $\|f\|_\infty := \sup_{x\in\mathbb{R}^m}|f(x)|$.

Let $(\Omega,\mathcal{F},\mathbb{P})$ be the probability space, where $\Omega := \{\omega : [0,T]\to\mathbb{R}^d;\ \omega \text{ is continuous},\ \omega(0)=0\}$, $\mathcal{F}$ is the Borel σ-field over Ω, and $\mathbb{P}:\mathcal{F}\to[0,1]$ is the Wiener measure on Ω. Let $W=\{W_t\}_t$ be a d-dimensional Brownian motion and let $\{\mathcal{F}_t\}_t$ be the filtration generated by W.

Let $T>0$. We consider the following N-dimensional SDE with a parameter $\lambda\in(0,1]$:

$$
\begin{aligned}
dX_s^{t,x,\lambda} &= b^\lambda(s,X_s^{t,x,\lambda})ds + \lambda\sum_{i=1}^d \sigma_i(s,X_s^{t,x,\lambda})dW_s^i,\quad X_t^{t,x,\lambda}\\
&= x\in\mathbb{R}^N,\quad 0\le t\le s\le T,
\end{aligned}
$$

where the map $(\lambda,s,y)\mapsto b^\lambda(s,y)$ belongs to $C_b^\infty([0,1]\times[0,T]\times\mathbb{R}^N;\mathbb{R}^N)$ and $\sigma_i\in C_b^\infty([0,T]\times\mathbb{R}^N;\mathbb{R}^N)$, $i=1,\ldots,d$. We assume the following condition: there exists $\epsilon>0$ such that $\sigma(t,x)\otimes\sigma(t,x)\ge\epsilon I_N$ for all $t\in[0,T]$ and $x\in\mathbb{R}^N$, where $I_N\in\mathbb{R}^{N\otimes N}$ is the identity matrix.

Let $K\in\mathbb{N}$ denote the number of the early-exercise dates of a Bermudan option and $\tau_j=Tj/K$, $j=0,1,\ldots,K$ denote each of the early-exercise date.

We define $f:\{\tau_0,\ldots,\tau_K\}\times\mathbb{R}^N\to\mathbb{R}$ as the discounted payoff function of a Bermudan option at each early-exercise date. We assume that the payoff f is a bounded measurable function. Here, the price of the Bermudan option is given by

$$
v_{\tau_K}(x) = f(\tau_K,x), \tag{1}
$$

$$
v_{\tau_j}(x) = (P_{\tau_j,\tau_{j+1}}v_{\tau_{j+1}})(x)\vee f(\tau_j,x),\quad j=K-1,\ldots,1,0, \tag{2}
$$

for $x\in\mathbb{R}^N$, where $p\vee q=\max\{p,q\}$ and

$$
P_{\tau_j,\tau_{j+1}}g(x) = E[g(X_{\tau_{j+1}}^{\tau_j,x,\lambda})],\quad x\in\mathbb{R}^N,\ j=0,1,\ldots,K-1, \tag{3}
$$

for a bounded measurable function $g:\mathbb{R}^N\to\mathbb{R}$.

The goal is to construct an efficient approximation for the Bermudan option price $v_0(x)$ for $x\in\mathbb{R}^N$.

3 Method

3.1 Sketch of Scheme

Before introducing the precise description of the algorithm, we describe a brief sketch of the deep learning-based algorithm for Bermudan option pricing based on asymptotic expansion of Wiener functionals. An efficient deep learning-based algorithm is constructed by the following two steps:

1. In the first step, let $m \in \mathbb{N}$ and we prepare operators $Q^{(m)}_{\tau_j,\tau_{j+1}}$ $j = n-1, \dots, 1, 0$, acting on a bounded measurable function $g : \mathbb{R}^N \to \mathbb{R}$ such that

$$\|(P_{\tau_j,\tau_{j+1}} - Q^{(m)}_{\tau_j,\tau_{j+1}})g\|_\infty = O(\lambda^{m+1} K^{-(m+1)/2}), \tag{4}$$

where the operators $Q^{(m)}_{\tau_j,\tau_{j+1}}$ $j = n-1, \dots, 1, 0$ are constructed based on asymptotic expansion of Wiener functionals. Then, the Bermudan option price $v_0(x)$ is approximated by

$$|v_0(x) - \bar{v}^{(m)}_0(x)| = O(\lambda^{m+1}), \tag{5}$$

where $\bar{v}_{\tau_j}$, $j = K, \dots, 1, 0$ is recursively defined by

$$\bar{v}^{(m)}_{\tau_K}(x) = f(\tau_K, x), \tag{6}$$

$$\bar{v}^{(m)}_{\tau_j}(x) = (Q^{(m)}_{\tau_j,\tau_{j+1}} \bar{v}^{(m)}_{\tau_{j+1}})(x) \vee f(\tau_j, x), \quad j = K-1, \dots, 1, 0, \tag{7}$$

for $x \in \mathbb{R}^N$.

2. In the second step, for $j = K-1, \dots, 1, 0$, we prepare neural networks $\mathcal{NN}^{\Theta^*_j}_{\tau_j}$ (we describe it precisely in Sect. 3.3) recursively defined by:

$$\bar{v}^{(m)}_{\tau_{K-1}} = Q^{(m)}_{\tau_{K-1},\tau_K} f(\tau_K, \cdot) \approx \mathcal{NN}^{\Theta^*_{K-1}}_{\tau_{K-1}}, \tag{8}$$

$$\begin{aligned} Q^{(m)}_{\tau_j,\tau_{j+1}} \bar{v}^{(m)}_{\tau_{j+1}} &\approx Q^{(m)}_{\tau_j,\tau_{j+1}} (\mathcal{NN}^{\Theta^*_{j+1}}_{\tau_{j+1}} \vee f(\tau_{j+1}, \cdot)) \\ &\approx \mathcal{NN}^{\Theta^*_j}_{\tau_j}, \quad j = K-2, \dots, 1, 0. \end{aligned} \tag{9}$$

Then we have a deep learning-based approximation for the Bermudan option price $v_o(x)$ as:

$$|v_0(x) - \bar{v}^{(m),\Theta^*_0}_0(x)| \;=\; O(\lambda^{m+1}),$$

where $\bar{v}^{(m),\Theta^*_0}_0(x) = (\mathcal{NN}^{\Theta^*_0}_0 \vee f)(0, x)$.

The sketch of the proposed scheme is described in Fig. 1.

In the following sections, we show our main result on the error estimate and the deep learning-based implementation for the method.

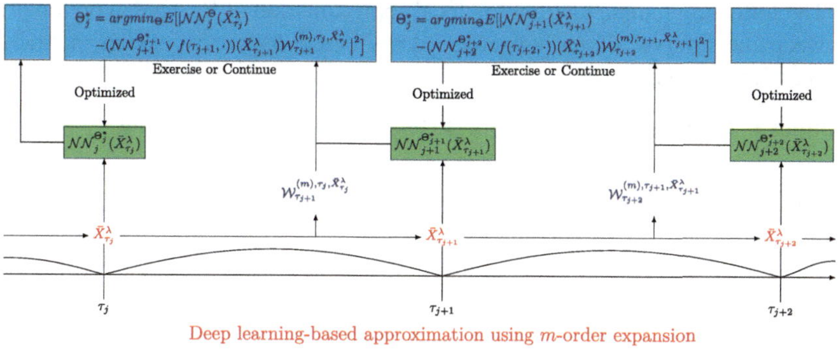

Fig. 1 Sketch of scheme

3.2 Expansion of Bermudan Option Price

Let H be the Cameron-Martin space, i.e. $H := \{h \in \Omega;\ \dot{h} \in L^2([0,T];\mathbb{R}^d)\}$ equipped with the inner product $\langle h_1, h_2\rangle_H := \int_0^T \dot{h}_1(s)\dot{h}_2(s)ds$. For $k \in \mathbb{N}$ and $p \in [1,\infty)$, let $\mathbb{D}^{k,p}$ be the Sobolev space of Wiener functionals with the norm $\|F\|_{k,p} = \|F\|_p + \sum_{i=1}^k \|D^i F\|_{L^p(\Omega;H^{\otimes i})}$, where D^i is the i-th Malliavin derivative operator for $i \leq k$. Let $\mathbb{D}^\infty = \bigcap_{p\geq 1}\bigcap_{k\in\mathbb{N}} \mathbb{D}^{k,p}$. For $F = (F^1,\ldots,F^N) \in (\mathbb{D}^\infty)^N$, the Malliavin covariance matrix $\sigma^F = [\sigma^F_{ij}]_{1\leq i,j\leq N}$ of F is defined by $\sigma^F_{ij} = \langle DF^i, DF^j\rangle_H$, $1 \leq i,j \leq N$. We say that $F \in (\mathbb{D}^\infty)^N$ is nondegenerate if σ^F is invertible a.s. and $\|\det(\sigma^F)^{-1}\|_p < \infty$ for all $p \geq 1$. Let $F \in (\mathbb{D}^\infty)^N$ be a nondegenerate Wiener functional and let $G \in \mathbb{D}^\infty$. For any $\varphi \in C_b^\infty(\mathbb{R}^N)$ and multi-index $\gamma = (\gamma_1,\ldots,\gamma_k) \in \{1,\ldots,N\}^k$, $k \in \mathbb{N}$, we have

$$E[\partial^\gamma \varphi(F)G] = E[\varphi(F)H_\gamma(F,G)], \tag{10}$$

where $\partial^\gamma\varphi(x) = \frac{\partial^k}{\partial x_{\gamma_1}\cdots\partial x_{\gamma_k}}\varphi(x)$ and $H_\gamma(F,G) = H_{(\gamma_k)}(F, H_{(\gamma_1,\ldots,\gamma_{k-1})}(F,G))$ with
$H_{(i)}(F,G) = \sum_{j=1}^N D^*([\sigma^F]^{-1}_{ij} DF^j G)$, $i = 1,\ldots,N$, where D^* is the adjoint operator of D. See [20, 28] for more details.

Under the condition on the diffusion process, for $m \geq 1$, $s > t$ and $x \in \mathbb{R}^N$, it holds that

$$\|X_s^{t,x,\lambda} - \{X_s^{t,x,0} + \lambda X_{1,s}^{t,x} + \cdots + \lambda^m X_{m,s}^{t,x}\}\|_{k,p} = O(\lambda^{m+1}), \tag{11}$$

for all $p > 1$ and $k \in \mathbb{N}$, where $X_{i,s}^{t,x} = \frac{1}{i!}\frac{\partial^i}{\partial\varepsilon^i}X_s^{t,x,\varepsilon}|_{\varepsilon=0}$, $i \in \mathbb{N}$. Define a scheme:

$$\bar{X}_s^{t,x,\lambda} = X_s^{t,x,0} + \lambda X_{1,s}^{t,x},\quad s > t,\ x \in \mathbb{R}^N, \tag{12}$$

where $X_\cdot^{t,x,0}$ is the solution to $\dot{X}_s^{t,x,0} = b^0(s, X_s^{t,x,0})$, $X_t^{t,x,0} = x$.

The following lemma justifies the local approximation of Bermudan option price between the interval of early-exercise dates based on an expansion method of Wiener functionals.

Lemma 1 *For $m \in \mathbb{N}$, there exists $C > 0$ independent of K such that*

$$\sup_{x\in\mathbb{R}^N} \left| E[g(X^{\tau_j,x,\lambda}_{\tau_{j+1}})] - E[g(\bar{X}^{\tau_j,x,\lambda}_{\tau_{j+1}})\mathcal{W}^{(m),\lambda,\tau_j,x}_{\tau_{j+1}}] \right| \leq C\|g\|_\infty K^{-(m+1)/2}\lambda^{m+1}, \tag{13}$$

for all bounded measurable functions $g : \mathbb{R}^N \to \mathbb{R}$, $\lambda \in (0, 1]$ and $j = 0, 1, \dots, K-1$, where

$$\mathcal{W}^{(m),\lambda,\tau_j,x}_{\tau_{j+1}} = 1 + \sum_{i=1}^{m} \lambda^i \sum_{k,\alpha^{(k)},\beta^{(k)},\gamma^{(p)}}^{(i)} H_{\alpha^{(k)}*\gamma^{(p)}}(X^{\tau_j,x}_{1,\tau_{j+1}}, 1)$$
$$\times \frac{1}{p!}\left\langle DX^{\tau_j,x,\gamma_1}_{1,\tau_{j+1}} \otimes \cdots \otimes DX^{\tau_j,x,\gamma_p}_{1,\tau_{j+1}}, E[D^p \prod_{e=1}^{k} X^{\tau_j,x,\alpha_e}_{1+\beta_e,\tau_{j+1}}]\right\rangle_{\mathcal{H}^{\otimes p}}, \tag{14}$$

with a convention:

$$\sum_{k,\alpha^{(k)},\beta^{(k)},\gamma^{(p)}}^{(i)} = \sum_{k=1}^{i} \sum_{\substack{\beta^{(k)}=(\beta_1,\dots,\beta_k),\\ \sum_\ell^k \beta_\ell=i}} \sum_{\alpha^{(k)}=(\alpha_1,\dots,\alpha_k)\in\{1,\dots,N\}^k} \frac{1}{k!} \sum_{\substack{\gamma^{(p)}=(\gamma_1,\dots,\gamma_p)\in\{1,\dots,N\}^p,\\ p\geq 0}} . \tag{15}$$

Proof See in Appendix 1: Proof of Lemma 1. □

Remark 1 The weight $\mathcal{W}^{(m),\lambda,\tau_j,x}_{\tau_{j+1}}$ is given by a polynomial of $X^{\tau_j,x}_{1,\tau_{j+1}}$.

Define $Q^{(m)}_{\tau_j,\tau_{j+1}}g(x) := E[g(\bar{X}^{\tau_j,x,\lambda}_{\tau_{j+1}})\mathcal{W}^{(m),\lambda,\tau_j,x}_{\tau_{j+1}}]$ for a bounded measurable function $g : \mathbb{R}^N \to \mathbb{R}$, $x \in \mathbb{R}^N$ and $j = 0, 1, \dots, K-1$. We construct an approximation for Bermudan option prices for the exercise dates:

$$\bar{v}^{(m)}_{\tau_K}(x) = f(\tau_K, x), \tag{16}$$

$$\bar{v}^{(m)}_{\tau_j}(x) = (Q^{(m)}_{\tau_j,\tau_{j+1}}\bar{v}^{(m)}_{\tau_{j+1}})(x) \vee f(\tau_j, x), \quad j = K-1, \dots, 1, 0. \tag{17}$$

The main result is given as follows.

Theorem 2 (m-Order Expansion for Bermudan Option Price v_0) *There exists $C > 0$ independent of K such that*

$$\sup_{x \in \mathbb{R}^N} \left| v_0(x) - \bar{v}_0^{(m)}(x) \right| \leq C \max_{1 \leq j \leq K} \| f(\tau_j, \cdot) \|_\infty \lambda^{m+1}, \tag{18}$$

for all $\lambda \in (0, 1]$.

Proof See Appendix 2: Proof of Theorem 2. □

3.3 Deep Learning Algorithm

We show a deep learning-based implementation for the approximations $Q^{(m)}_{\tau_j,\tau_{j+1}} \bar{v}^{(m)}_{\tau_{j+1}}$ and $\bar{v}^{(m)}_{\tau_j}$ for $j = K-1, \ldots, 1, 0$. By the property of conditional expectation, the function $Q^{(m)}_{\tau_j,\tau_{j+1}} \bar{v}^{(m)}_{\tau_{j+1}}$ satisfies that

$$\begin{aligned} Q^{(m)}_{\tau_j,\tau_{j+1}} \bar{v}^{(m)}_{\tau_{j+1}} = argmin_{\phi \in L^2(\mathbb{R}^N, \mathscr{B}(\mathbb{R}^N), P \circ (\bar{X}^{\tau_0,x,\lambda}_{\tau_j})^{-1})} E\Big[\Big| \phi(\bar{X}^{\tau_0,x,\lambda}_{\tau_j}) \\ - (Q^{(m)}_{\tau_{j+1},\tau_{j+2}} \bar{v}^{(m)}_{\tau_{j+2}} \vee f(\tau_{j+1}, \cdot))(\bar{X}^{\tau_0,x,\lambda}_{\tau_{j+1}}) \mathcal{W}^{(m),\lambda,\tau_j,\bar{X}^{\tau_0,x,\lambda}_{\tau_j}}_{\tau_{j+1}} \Big|^2 \Big], \end{aligned} \tag{19}$$

where $Q^{(m)}_{\tau_{j+1},\tau_{j+2}} \bar{v}^{(m)}_{\tau_{j+2}} = f(\tau_{j+1}, \cdot)$ if $j = K-1$. Then, we solve the minimization problem (19) by deep learning.

Now, we describe the deep learning-based algorithm of the proposed scheme. Let $\nu \in \mathbb{N}$. For $p, \ell \in \mathbb{N}$, $e \in \{0\} \cup \mathbb{N}$ and $\theta = (\theta^1, \ldots, \theta^\nu) \in \mathbb{R}^\nu$ such that $q + \ell(p+1) \leq \nu$, let $A^{\theta,q}_{p,\ell} : \mathbb{R}^p \to \mathbb{R}^\ell$ denote an affine transformation function given by

$$A^{\theta,q}_{p,\ell}(x) = \begin{pmatrix} \theta^{q+1} & \cdots & \theta^{q+p} \\ \vdots & \ddots & \vdots \\ \theta^{q+(\ell-1)p+1} & \cdots & \theta^{q+\ell p} \end{pmatrix} \begin{pmatrix} x_1 \\ \vdots \\ x_p \end{pmatrix} + \begin{pmatrix} \theta^{q+\ell p+1} \\ \vdots \\ \theta^{q+\ell p+\ell} \end{pmatrix}, \quad x \in \mathbb{R}^p.$$

Let $\mathcal{L}_r : \mathbb{R}^r \to \mathbb{R}$ denote the Rectified Linear Unit (ReLU) activation function given by

$$\mathcal{L}_r(x) = (\max\{x_1, 0\}, \ldots, \max\{x_r, 0\}), \quad x \in \mathbb{R}^r. \tag{20}$$

Then, neural networks are represented by the composition of the affine transformation function and the ReLU function.

Let $s \in \{3,4,\ldots\}$ and $d_0, d_1, \ldots, d_s \in \mathbb{N}$ such that $\sum_{i=1}^{s} d_i(d_{i-1}+1) = \nu$. Here, we have a deep learning-based approximation $Q^{(m),\mathcal{NN},\theta^*_{\tau_j}}_{\tau_j,\tau_{j+1}} \bar{v}^{(m)}_{\tau_{j+1}}$ for the function $Q^{(m)}_{\tau_j,\tau_{j+1}} \bar{v}^{(m)}_{\tau_{j+1}}$ with the optimized parameter $\theta^*_{\tau_j} \in \mathbb{R}^\nu$ and an approximation:

$$\bar{v}^{(m)}_{\tau_j} \approx Q^{(m),\mathcal{NN},\theta^*_{\tau_j}}_{\tau_j,\tau_{j+1}} \bar{v}^{(m)}_{\tau_{j+1}} \vee f(\tau_j,\cdot)$$

where $Q^{(m),\mathcal{NN},\theta^*_{\tau_j}}_{\tau_j,\tau_{j+1}} \bar{v}^{(m)}_{\tau_{j+1}}$ is a deep neural network, composed of $(s+1)$-th layers (1 input layer with d_0-neurons, $s-1$-hidden layers with d_i-neurons for each i-th layer and 1 output layer with d_s-neurons), given by

$$Q^{(m),\mathcal{NN},\theta}_{\tau_j,\tau_{j+1}} \bar{v}^{(m)}_{\tau_{j+1}}(x) = (A^{\theta,\sum_{i=1}^{s-1} d_i(d_{i-1}+1)}_{d_{s-1},1} \circ \mathcal{L}_{d_{s-1}} \circ A^{\theta,\sum_{i=1}^{s-2} d_i(d_{i-1}+1)}_{d_{s-2},d_{s-1}}$$
$$\circ \cdots \circ \mathcal{L}_{d_2} \circ A^{\theta,d_1(d_0+1)}_{d_1,d_2} \circ \mathcal{L}_{d_1} \circ A^{\theta,0}_{d_0,d_1})(x), \ x \in \mathbb{R}^N$$

with the optimized parameter $\theta^*_{\tau_j}$ which satisfies $\theta^*_{\tau_j} = argmin_\theta L_{\tau_j}(\theta)$ with the loss function $L_{\tau_j} : \mathbb{R}^\nu \to \mathbb{R}$ defined by:

$$L_{\tau_j}(\theta) = E\Big[\Big| Q^{(m),\mathcal{NN},\theta}_{\tau_j,\tau_{j+1}} \bar{v}^{(m)}_{\tau_{j+1}} (\bar{X}^{\tau_0,x,\lambda}_{\tau_j})$$
$$- (Q^{(m),\mathcal{NN},\theta^*_{\tau_{j+1}}}_{\tau_{j+1},\tau_{j+2}} \bar{v}^{(m)}_{\tau_{j+2}} \vee f(\tau_{j+1},\cdot))(\bar{X}^{\tau_0,x,\lambda}_{\tau_{j+1}}) \mathcal{W}^{(m),\lambda,\tau_j,\bar{X}^{\tau_0,x,\lambda}_{\tau_{j+1}}}_{\tau_{j+1}} \Big|^2\Big].$$

We show the sketch of the algorithm in Algorithm 1 in the next page.

In the next section, we show numerical results for the deep learning-based approximation in order to check the accuracy and efficiency of the proposed algorithm.

4 Numerical Examples

In order to check the validity of the proposed method, we provide some numerical examples in this section. Each of the numerical experiments were performed in Python using Tensorflow 1.15 on two GPUs (NVIDIA GA102GL [RTX A6000] 48GB with NVLink) where the underlying system consists of AMD EPYC 7402P 2.80GHz 24 cores CPU with 128 GB DDR4-3200 memory running on Ubuntu 18.04.

4.1 5-Dimensional Black-Scholes Model

For the first example, we consider the case of Bermudan maximum call option under the 5-dimensional Black–Scholes model:

Algorithm 1 Deep learning-based expansion for pricing Bermudan option

Require: $M \in \mathbb{N}$ (batch size), $J \in \mathbb{N}$ (number of train-steps), $\gamma \in (0, 1)$ (learning rate), $K \in \mathbb{N}$ (number of exercise dates), $n \in \mathbb{N}$ (number of time steps for each interval of the early-exercise dates)

for $i = K - 1$ to 0 **do**
  for $j = 1$ to J **do**
    for $r = 1$ to M **do**
      $\bar{X}_{\tau_0}^{\tau_0,x,\lambda,r,j} = x,$
      for $\ell = 1$ to $(i + 1)$ **do**
        Generate i.i.d. Gaussian RVs $\Delta W_{t_{\ell-1},t_\ell}^{1,r,j}, \ldots, \Delta W_{t_{\ell-1},t_\ell}^{d,r,j}$
        $\bar{X}_{\tau_\ell}^{\tau_0,x,\lambda,r,j} = \bar{X}_{\tau_{\ell-1}}^{\tau_0,x,\lambda,r,j} + b^\lambda(\tau_{\ell-1}, \bar{X}_{\tau_{\ell-1}}^{\tau_0,x,\lambda,r,j})(\tau_{\ell-1}, \tau_\ell)$
        $+\lambda \sum_{e=1}^{d} \sigma_e(\tau_{\ell-1}, \bar{X}_{\tau_{\ell-1}}^{\tau_0,x,\lambda,r,j})\Delta W_{\tau_{\ell-1},\tau_\ell}^{e,r,j}$
      end for
      Compute $\mathcal{W}_{\tau_\ell}^{(m),\tau_{\ell-1},\bar{X}_{\tau_{\ell-1}}^{\tau_0,x,\lambda,r,j}}$
    end for
    if $i = K - 1$ **then**
      $Q_{\tau_{i+1},\tau_{i+2}}^{(m),\mathcal{NN},\Theta_{\tau_{i+1}}} \bar{v}_{\tau_{i+2}}^{(m)}(y) = f(\tau_{i+1}, y)$
    end if
    $L_{\tau_i,j}(\theta_{\tau_i}^j) = \frac{1}{M} \sum_{r=1}^{M} \Big[Q_{\tau_i,\tau_{i+1}}^{(m),\mathcal{NN},\theta_{\tau_i}^j} \bar{v}_{\tau_{i+1}}^{(m)}(\bar{X}_{\tau_i}^{\tau_0,x,\lambda,r,j})$
    $-(Q_{\tau_{i+1},\tau_{i+2}}^{(m),\mathcal{NN},\Theta_{\tau_{i+1}}} \bar{v}_{\tau_{i+2}}^{(m)} \vee f(\tau_{i+1}, \cdot))(\bar{X}_{\tau_{i+1}}^{\tau_0,x,\lambda,r,j})\mathcal{W}_{\tau_{i+1}}^{(m),\lambda,\tau_i,\bar{X}_{\tau_i}^{\tau_0,x,\lambda,r,j}} \Big]^2,$
    where $Q_{\tau_i,\tau_{i+1}}^{(m),\mathcal{NN},\theta} \bar{v}_{\tau_{i+1}}^{(m)}$ is a function given by neural network
    Minimize the loss function $L_{\tau_i,j}(\theta_{\tau_i}^j)$ and obtain the parameter $\theta_{\tau_i}^{*j}$ using the ADAM optimizer [15]
    Update the parameter $\theta_{\tau_i}^{j+1} = \theta_{\tau_i}^{*j}$
  end for
  $\Theta_{\tau_i} = \theta_{\tau_i}^J$
end for
Return $(Q_{\tau_0,\tau_1}^{(m),\mathcal{NN},\Theta_{\tau_0}} \bar{v}_{\tau_1}^{(m)} \vee f(\tau_0, \cdot))(x)$

$$dX_t^{i,\lambda} = (r - q)X_t^{i,\lambda}dt + \lambda X_t^{i,\lambda}dB_t^i, \tag{21}$$

for $i = 1, \ldots, 5$ and $X_0^\lambda = x \in \mathbb{R}^5$. We set the parameters as $K = 9$, $T = 3.0$, $x = (100.0, \ldots, 100.0)$, $r = 0.05$, $q = 0.1$, $\lambda = 0.2$ and the payoff function as $f : [0, T] \times \mathbb{R}^5 \to \mathbb{R}$ as $f(t, x) = e^{-rt}(0.0 \vee (80.0 - x_1) \vee \cdots \vee (80.0 - x_5))$.

The deep neural network used in this section is composed of 1-input layer with 5-neurons, 2-hidden layers with 15-neurons each and 1-output layer with 1-neuron to implement the proposed scheme. We set the hyperparameters of deep learning as the batch size $M = 32768$, the train steps $J = 2000$ and the learning rate $\gamma(j) = 2.5 \times 10^{-1}\mathbf{1}_{[0,0.3J]}(j) + 2.5 \times 10^{-2}\mathbf{1}_{(0.3J,0.6J]}(j) + 2.5 \times 10^{-3}\mathbf{1}_{(0.6J,J]}(j)$, $j \leq J$. In the implementation, we employ the ADAM optimizer [15] and batch normalization [14]. We compute the algorithms with $m = 1$ and $m = 2$, i.e. the expansions with $O(\lambda^2)$ and $O(\lambda^3)$, which are referred to as "Deep AE 2nd" and "Deep AE 3rd", respectively.

Table 1 Numerical result for 5-dimensional Black-Scholes model

Scheme	Rel. err.	Std. dev.	Avg. comp. time
Deep EM ($n = 2^3$)	0.00584	0.0674	102.39 s
Deep EM ($n = 2^6$)	0.00162	0.1013	316.41 s
Deep AE 2nd	0.00334	0.0822	68.94 s
Deep AE 3rd	0.00032	0.0807	72.55 s
DOS ($n = 2^6$)	0.00485	0.0645	103.96 s

For comparison, we discretize the intervals of early-exercise dates n-times by the Euler-Maruyama scheme and compute the Bermudan option price with two types of deep learning algorithms. One is the same deep learning algorithm as the proposed method (referred to as "Deep EM") and the other is an algorithm called the deep optimal stopping introduced in [4] (referred to as "DOS"). On the other hand, the proposed asymptotic expansion method does not require any time-discretization between the intervals of the early-exercise dates. In this sense, the proposed method is implemented with a one-shot discretization ($n = 1$).

Neural networks used in the implementation of DOS are the same settings as the proposed method, except for the input layer. The neurons of the input layer of DOS is set as $(5 + 1)$ by adding the payoff to the input for each neural network according to [4]. In other words, we employ $(5 + 1)$-dimensional input $(\bar{X}_{\tau_j}^{\tau_0,x,\lambda}, f(\tau_j, \bar{X}_{\tau_j}^{\tau_0,x,\lambda}))_{j=0,\dots,N}$ to implement DOS.

The approximation error for each scheme is estimated by $|v_0^{\mathrm{Ref}}(x) - v_0^{\text{Deep AE 2nd}}(x)|/v_0^{\mathrm{Ref}}(x)$, $|v_0^{\mathrm{Ref}}(x) - v_0^{\text{Deep AE 3rd}}(x)|/v_0^{\mathrm{Ref}}(x)$, $|v_0^{\mathrm{Ref}}(x) - v_0^{\text{Deep EM},n}(x)|/v_0^{\mathrm{Ref}}(x)$ and $|v_0^{\mathrm{Ref}}(x) - v_0^{\mathrm{DOS},n}(x)|/v_0^{\mathrm{Ref}}(x)$, where $v_0^{\text{Deep AE 2nd}}(x)$, $v_0^{\text{Deep AE 3rd}}(x)$ are obtained by the average of 25 independent trials of each method and $v_0^{\text{Deep EM},n}(x)$ and $v_0^{\mathrm{DOS},n}(x)$ are computed in the same manner for each number of time steps n. The reference value $v_0^{\mathrm{Ref}}(x)$ is obtained by 25 independent trials of the proposed deep learning-based algorithm with the explicit solution of X obtained by the Itô formula. The numerical result is summarized in Table 1. The table shows compared to Deep EM and DOS, the proposed methods provide more accurate result efficiently with shorter computing time. Moreover, as the order of asymptotic expansion grows, the approximation error decreases, which is consistent with our theoretical result.

4.2 100-*Dimensional SABR Model (*50-*Assets and* 50-*Volatility Processes)*

We show the example under a higher dimensional setting where it cannot be possible to compute by the classical algorithms for pricing Bermudan options. We consider the following 100-dimensional SABR model which describes the dynamics of 50-assets with 50-stochastic volatility processes:

Table 2 100-dimensional SABR model

Scheme	Rel. err.	Std. dev.	Avg. comp. time
Deep EM ($n = 2^3$)	0.00222	0.0308	112.92 s
Deep EM ($n = 2^5$)	0.00048	0.0314	219.51 s
Deep AE 2nd	0.00158	0.0212	73.49 s
Deep AE 3rd	0.00036	0.0224	100.06 s
DOS ($n = 2^5$)	0.00193	0.0673	127.01 s

$$\begin{cases} dX_t^{2i-1,\lambda} = \lambda C X_t^{2i-1,\lambda} X_t^{2i,\lambda} dB_t^{2i-1}, \\ dX_t^{2i,\lambda} = \lambda X_t^{2i,\lambda}(\rho dB_t^{2i-1} + \sqrt{1-\rho^2} dB_t^{2i}), \end{cases} \tag{22}$$

for $i = 1, \ldots, 50$, and $X_0^\lambda = x \in \mathbb{R}^{100}$. In this section, we compute the price of the Bermudan maximum put option under this model (22).

The parameters are set as $K = 4$, $T = 1.0$, $(x_1, x_3, \ldots, x_{99}) = (100.0, \ldots, 100.0)$, $(x_2, x_4, \ldots, x_{100}) = (0.1, \ldots, 0.1)$, $\lambda = 0.1$, $C = 10$ and $\rho = -0.5$. We assume the zero risk free rate and set the payoff function f as $f(x) = 0.0 \vee (90.0 - x_1) \vee (90.0 - x_3) \vee \cdots \vee (90.0 - x_{99})$.

We prepare the deep neural network composed of 1-input layer with 100-neurons, 2-hidden layers with 110-neurons each and 1-output layer with 1-neuron. The hyperparameters are set as the batch size $M = 4096$ and the train steps $J = 5000$ and use the learning rate $\gamma(j) = 5.0 \times 10^{-1}\mathbf{1}_{[0,0.3J]}(j) + 5.0 \times 10^{-2}\mathbf{1}_{(0.3J,0.6J]}(j) + 5.0 \times 10^{-3}\mathbf{1}_{(0.6J,J]}(j)$, $j \leq J$. The ADAM optimizer [15] and batch normalization [14] are utilized. We compute the proposed algorithm Deep AE 2nd and Deep AE 3rd in the numerical experiment.

Also we compare the result of Deep EM and DOS for each number of time steps n. The algorithm Deep EM is implemented using the same neural network settings as Deep AE 2nd. The algorithm DOS is also implemented using the same neural network except for the input layer. The neurons of the input layer of DOS is set as $(100 + 1)$ by adding the payoff to the input for each neural network according to [4].

The error of each scheme is estimated in the same manner as the previous example. The reference value is computed by the average of 25-independent trials of the first order method with $n = 2^8$. The numerical result is summarized in Table 2.

From the table, the proposed algorithms achieve efficient computing results compared to Deep EM an DOS in terms of the approximation error, the standard deviation and the computing time. Moreover, we confirm that increasing the order of asymptotic expansion leads to more accurate computing results, consistently with the theoretical rate of convergence.

Throughout the numerical experiments, we can check the validity of the proposed method, even in a high dimensional setting where the classical algorithm is difficult to compute.

5 Concluding Remarks

We introduced a new deep learning-based algorithm for high-dimensional Bermudan option pricing based on asymptotic expansion of Wiener functionals. The proposed method is efficiently implemented in a high-dimensional setting where the classical algorithms such as the least squares Monte Carlo method cannot work. Moreover, we confirmed that it achieves accurate approximations for Bermudan option prices both theoretically and numerically.

For future work, it is worth considering to investigate the collaboration between other deep learning-based algorithms such as the deep optimal stopping of [4] with expansion method of Wiener functionals or weak approximation of SDEs.

Acknowledgments This work was supported by JST PRESTO (JPMJPR2029), Japan.

Appendix 1: Proof of Lemma 1

We apply the argument of [25]. Let $F_{\tau_{j+1}}^{\tau_j,x,\lambda} = (X_{\tau_{j+1}}^{\tau_j,x,\lambda} - X_{\tau_{j+1}}^{\tau_j,x,0})/\lambda$, $\lambda \in (0,1]$. Note that there exists $C > 0$ such that

$$\sup_{\xi\in[0,1]} \left| \det \sigma^{F_{\tau_{j+1}}^{\tau_j,x,0} + \xi(F_s^{t,x,\varepsilon} - F_{\tau_{j+1}}^{\tau_j,x,0})} - \det \sigma^{F_{\tau_{j+1}}^{\tau_j,x,0}} \right|$$
$$\leq C \| D(F_{\tau_{j+1}}^{\tau_j,x,\lambda} - F_{\tau_{j+1}}^{\tau_j,x,0}) \|_H (\| DF_{\tau_{j+1}}^{\tau_j,x,0} \|_H + \| DF_{\tau_{j+1}}^{\tau_j,x,\lambda} \|_H)^{2d-1}.$$

We define

$$\eta_{\tau_{j+1}}^{\tau_j,x,\lambda} = C \| D(F_{\tau_{j+1}}^{\tau_j,x,\lambda} - F_{\tau_{j+1}}^{\tau_j,x,0}) \|_H (\| DF_{\tau_{j+1}}^{\tau_j,x,0} \|_H$$
$$+ \| DF_{\tau_{j+1}}^{\tau_j,x,\lambda} \|_H)^{2d-1} / \det \sigma^{F_{\tau_{j+1}}^{\tau_j,x,0}}$$

and $\psi(x) = \mathbf{1}_{|x|\leq 1/4} + \exp(1 - (1/4)^2/((1/4)^2 - (x-1/4)^2))\mathbf{1}_{1/4<|x|<1/2}$, $x \in \mathbb{R}$. Note that $\psi(\eta_{\tau_{j+1}}^{\tau_j,x,\lambda}) \neq 0$ implies $\det \sigma^{F_{\tau_{j+1}}^{\tau_j,x,0} + \xi(F_{\tau_{j+1}}^{\tau_j,x,\lambda} - F_{\tau_{j+1}}^{\tau_j,x,0})} \geq \frac{1}{2} \det \sigma^{F_{\tau_{j+1}}^{\tau_j,x,0}}$ for all $\xi \in [0,1]$. Also, it holds that for all $k \in \mathbb{N}$, $p > 1$, $\|\psi(\eta_{\tau_{j+1}}^{\tau_j,x,\lambda})\|_{k,p} = O(1)$ and $\|1 - \psi(\eta_{\tau_{j+1}}^{\tau_j,x,\lambda})\|_{k,p} = O(\lambda^r K^{-r/2})$ for arbitrary $r > 1$. Let $g : \mathbb{R}^N \to \mathbb{R}$ be a bounded Borel measurable function and consider the decomposition $E[g(X_{\tau_{j+1}}^{\tau_j,x,\lambda})] = E[g(X_{\tau_{j+1}}^{\tau_j,x,\lambda})\psi(\eta_{\tau_{j+1}}^{\tau_j,x,\lambda})] + E[g(X_{\tau_{j+1}}^{\tau_j,x,\lambda})\{1 - \psi(\eta_{\tau_{j+1}}^{\tau_j,x,\lambda})\}]$. The second term of the right-hand side can be estimated as $\sup_x |E[g(X_{\tau_{j+1}}^{\tau_j,x,\lambda})\{1 - \psi(\eta_{\tau_{j+1}}^{\tau_j,x,\lambda})\}]| \leq \|g\|_\infty \lambda^{m+1} K^{-(m+1)/2}$. The first term is expanded

$$
\begin{aligned}
&E[g(X_{\tau_{j+1}}^{\tau_j,x,\lambda})\psi(\eta_{\tau_{j+1}}^{\tau_j,x,\lambda})] \\
&\quad = \int_{\mathbb{R}^N} g(X_{\tau_{j+1}}^{\tau_j,x,0} + \lambda y) E[\delta_y(F_{\tau_{j+1}}^{\tau_j,x,0}) \\
&\qquad \{1 + \sum_{i=1}^{m} \lambda^i \sum_{k,\alpha^{(k)},\beta^{(k)}}^{(i)} H_\alpha(F_{\tau_{j+1}}^{\tau_j,x,0}, \prod_{i=1}^{k} F_{\beta_i,\tau_{j+1}}^{\tau_j,x,\alpha_i})\}]dy \\
&\quad + R^{1,g}(\tau_j,\tau_{j+1},x,\lambda) + R^{2,g}(\tau_j,\tau_{j+1},x,\lambda)
\end{aligned} \tag{23}
$$

where $\sum_{k,\alpha^{(k)},\beta^{(k)}}^{(i)} = \sum_{k=1}^{i} \sum_{\beta^{(k)}=(\beta_1,\dots,\beta_k),\sum_\ell^k \beta_\ell = i} \sum_{\alpha^{(k)}=(\alpha_1,\dots,\alpha_k)\in\{1,\dots,N\}^k} \frac{1}{k!}$ and

$$
\begin{aligned}
&R^{1,g}(\tau_j,\tau_{j+1},x,\lambda) \\
&= \int_0^1 \frac{(1-\xi)^m}{m!} E\Big[g(X_{\tau_{j+1}}^{\tau_j,x,0} + \lambda \widetilde{F_{\tau_{j+1}}^{\tau_j,x,\xi,\lambda}}) \\
&\quad \sum_{\alpha\in\{1,\dots,N\}^{m+1}} H_\alpha\Big(\widetilde{F_{\tau_{j+1}}^{\tau_j,x,\xi,\lambda}}, \prod_{\ell=1}^{m+1}(F_{\tau_{j+1}}^{\tau_j,x,\lambda,\alpha_\ell} - F_{\tau_{j+1}}^{\tau_j,x,0,\alpha_\ell})\psi(\eta_{\tau_{j+1}}^{\tau_j,x,\lambda})\Big)\Big]d\xi
\end{aligned}
$$

with $\widetilde{F_{\tau_{j+1}}^{\tau_j,x,\xi,\lambda}} = F_{\tau_{j+1}}^{\tau_j,x,0} + \xi(F_{\tau_{j+1}}^{\tau_j,x,\lambda} - F_{\tau_{j+1}}^{\tau_j,x,0})$, $\xi \in [0,1]$ and

$$
\begin{aligned}
&R^{2,g}(\tau_j,\tau_{j+1},x,\lambda) \\
&= \sum_{\alpha\in\{1,\dots,N\}^k,k\le m} E[g(X_{\tau_{j+1}}^{\tau_j,x,0} + \lambda F_{\tau_{j+1}}^{\tau_j,x,0}) H_\alpha(F_{\tau_{j+1}}^{\tau_j,x,0}, G_{\alpha,\tau_{j+1}}^{\tau_j,x,\lambda}\psi(\eta_{\tau_{j+1}}^{\tau_j,x,\lambda}))] \\
&+ \sum_{\alpha\in\{1,\dots,N\}^k,k\le m} E[g(X_{\tau_{j+1}}^{\tau_j,x,0} + \lambda F_{\tau_{j+1}}^{\tau_j,x,0}) H_\alpha(F_{\tau_{j+1}}^{\tau_j,x,0}, \hat{G}_{\alpha,\tau_{j+1}}^{\tau_j,x,\lambda}(1-\psi(\eta_{\tau_{j+1}}^{\tau_j,x,\lambda})))]
\end{aligned}
$$

with $\{G_{\alpha,\tau_{j+1}}^{\tau_j,x,\lambda}\}_{\alpha\in\{1,\dots,N\}^k,k\le m,\lambda\in(0,1]}, \{\hat{G}_{\alpha,\tau_{j+1}}^{\tau_j,x,\lambda}\}_{\alpha\in\{1,\dots,N\}^k,k\le m,\lambda\in(0,1]} \subset \mathbb{D}^\infty$ such that for all $k \le m$ and $\alpha \in \{1,\dots,d\}^k$, $G_{\alpha,\tau_{j+1}}^{\tau_j,x,\lambda}$, $\hat{G}_{\alpha,\tau_{j+1}}^{\tau_j,x,\lambda}$, $\lambda \in (0,1]$ satisfy for all $\ell \in \mathbb{N}$, $p > 1$, $\|G_{\alpha,\tau_{j+1}}^{\tau_j,x,\lambda}\|_{\ell,p} = O(\lambda^{m+1}K^{-(m+1)/2})$, $\|\hat{G}_{\alpha,\tau_{j+1}}^{\tau_j,x,\lambda}\|_{\ell,p} = O(\lambda K^{-1/2})$. Here, for $\alpha \in \{1,\dots,N\}^k$ and $p \ge 1$, it holds that

$$
\begin{aligned}
&\Big\| H_\alpha\Big(\widetilde{F_{\tau_{j+1}}^{\tau_j,x,\xi,\lambda}}, \prod_{\ell=1}^{m+1}(F_{\tau_{j+1}}^{\tau_j,x,\lambda,\alpha_\ell} - F_{\tau_{j+1}}^{\tau_j,x,0,\alpha_\ell})\psi(\eta_{\tau_{j+1}}^{\tau_j,x,\lambda})\Big)\Big\|_p \\
&\quad \le C\|(\det\sigma^{F_{\tau_{j+1}}^{\tau_j,x,0}})^{-1}\|_{p_1}^{e} \|D\widetilde{F_{\tau_{j+1}}^{\tau_j,x,\xi,\lambda}}\|_{k,p_2,H}^{2Ne-|\alpha|} \Big\|
\end{aligned}
$$

$$\prod_{\ell=1}^{m+1}(F_{\tau_{j+1}}^{\tau_j,x,\lambda,\alpha_\ell}-F_{\tau_{j+1}}^{\tau_j,x,0,\alpha_\ell})\psi(\eta_{\tau_{j+1}}^{\tau_j,x,\lambda})\Big\|_{k,p_3}$$
$$=O(\lambda^{m+1}K^{-(m+1)/2})$$

for some $C>0$ and $e,p_1,p_2,p_3>1$, where we used $(\det\sigma^{\widetilde{F_{\tau_{j+1}}^{\tau_j,x,\xi,\lambda}}})^{-1}\leq 2(\det\sigma^{F_{\tau_{j+1}}^{\tau_j,x,0}})^{-1}$ on $\psi(\eta_{\tau_{j+1}}^{\tau_j,x,\lambda})\neq 0$. Also, we have

$$\begin{aligned}&\|H_\alpha(F_{\tau_{j+1}}^{\tau_j,x,0},G_{\alpha,s}^{\tau_j,x,\lambda}\psi(\eta_{\tau_{j+1}}^{\tau_j,x,\lambda}))\|_p\\ \leq&C\|(\det\sigma^{F_{\tau_{j+1}}^{\tau_j,x,0}})^{-1}\|_{p_1}^e\|DF_{\tau_{j+1}}^{\tau_j,x,0}\|_{k,p_2,H}^{2Ne-|\alpha|}\|G_{\alpha,\tau_{j+1}}^{\tau_j,x,\lambda}\psi(\eta_{\tau_{j+1}}^{\tau_j,x,\lambda})\|_{k,p_3}\\ =&O(\lambda^{m+1}K^{-(m+1)/2})\end{aligned}$$

and

$$\begin{aligned}&\|H_\alpha(F_{\tau_{j+1}}^{\tau_j,x,0},\hat{G}_{\alpha,s}^{\tau_j,x,\lambda}(1-\psi(\eta_{\tau_{j+1}}^{\tau_j,x,\lambda})))\|_p\\ \leq&C\|(\det\sigma^{F_{\tau_{j+1}}^{\tau_j,x,0}})^{-1}\|_{p_1}^e\|DF_{\tau_{j+1}}^{\tau_j,x,0}\|_{k,p_2,H}^{2Ne-|\alpha|}\|\hat{G}_{\alpha,\tau_{j+1}}^{\tau_j,x,\lambda}(1-\psi(\eta_{\tau_{j+1}}^{\tau_j,x,\lambda}))\|_{k,p_3}\\ =&O(\lambda^{m+1}K^{-(m+1)/2}).\end{aligned}$$

Therefore, there exists $C>0$ such that

$$\begin{aligned}\|R^{1,g}(\tau_j,\tau_{j+1},\cdot,\lambda)\|_\infty&\leq C\|g\|_\infty\lambda^{m+1}K^{-(m+1)/2},\\ \|R^{2,g}(\tau_j,\tau_{j+1},\cdot,\lambda)\|_\infty&\leq C\|g\|_\infty\lambda^{m+1}K^{-(m+1)/2},\end{aligned}$$

for all bounded measurable function $g:\mathbb{R}^N\to\mathbb{R}$, $\lambda\in(0,1]$ and $j=0,1,\ldots,K-1$. The expansion terms can be calculated as follows: for $G\in\mathbb{D}^\infty$,

$$\begin{aligned}&E[\partial^\alpha\delta_y(F_{\tau_{j+1}}^{\tau_j,x,0})G]\\ &=E[\partial^\alpha\delta_y(F_{\tau_{j+1}}^{\tau_j,x,0})]E[G]+E[\partial^\alpha\delta_y(F_{\tau_{j+1}}^{\tau_j,x,0})\sum_{p\geq 1}\frac{1}{p!}D^{*p}(E[D^pG])]\\ &=E[\partial^\alpha\delta_y(F_{\tau_{j+1}}^{\tau_j,x,0})]E[G]+\sum_{p\geq 1}\frac{1}{p!}E[\langle D^p\partial^\alpha\delta_y(F_{\tau_{j+1}}^{\tau_j,x,0}),E[D^pG]\rangle_{H^{\otimes p}}]\\ &=E[\partial^\alpha\delta_y(F_{\tau_{j+1}}^{\tau_j,x,0})]E[G]\\ &+\sum_{p\geq 1}\sum_{\gamma}E[\partial^{\alpha*\gamma}\delta_y(F_{\tau_{j+1}}^{\tau_j,x,0})\end{aligned}$$

$$\frac{1}{p!}\langle DF_{\tau_{j+1}}^{\tau_j,x,0,\gamma_1}\otimes\cdots\otimes DF_{\tau_{j+1}}^{\tau_j,x,0,\gamma_p},\, E[D^pG]\rangle_{H^{\otimes p}}]$$

$$=\sum_{p\geq 0}\sum_{\gamma} E[\delta_y(F_{\tau_{j+1}}^{\tau_j,x,0})H_{\alpha*\gamma}(F_{\tau_{j+1}}^{\tau_j,x,0},$$

$$\frac{1}{p!}\langle DF_{\tau_{j+1}}^{\tau_j,x,0,\gamma_1}\otimes\cdots\otimes DF_{\tau_{j+1}}^{\tau_j,x,0,\gamma_p},\, E[D^pG]\rangle_{H^{\otimes p}})].$$

Then, we have proved

$$\Big|E[g(X_{\tau_{j+1}}^{\tau_j,x,\lambda})]-\{E[g(\bar{X}_{\tau_{j+1}}^{\tau_j,x,\lambda})]+\sum_{i=1}^{m}\lambda^i\sum_{k,\alpha^{(k)},\beta^{(k)},\gamma^{(p)}}^{(i)}$$
$$E[g(\bar{X}_{\tau_{j+1}}^{\tau_j,x,\lambda})H_{\alpha^{(k)}*\gamma^{(p)}}(X_{1,\tau_{j+1}}^{\tau_j,x},1)]$$
$$\times\frac{1}{p!}\Big\langle DX_{1,\tau_{j+1}}^{\tau_j,x,\gamma_1}\otimes\cdots\otimes DX_{1,\tau_{j+1}}^{\tau_j,x,\gamma_p},\, E[D^p\prod_{e=1}^{k}X_{1+\beta_e,\tau_{j+1}}^{\tau_j,x,\alpha_e}]\Big\rangle_{H^{\otimes p}}\}\Big|$$
$$\leq C\|g\|_\infty K^{-(m+1)/2}\lambda^{m+1}, \tag{24}$$

for all bounded measurable functions $g:\mathbb{R}^N\to\mathbb{R}$, $x\in\mathbb{R}^N$, $\lambda\in(0,1]$ and $j=0,1,\ldots,K-1$. □

Appendix 2: Proof of Theorem 2

We write the definitions of $v_0(x)$ and $\bar{v}_0^{(m)}(x)$ again:

$$v_0(x)=P_{\tau_0,\tau_1}v_{\tau_1}(x)\vee f(\tau_0,x) \tag{25}$$
$$\bar{v}_0^{(m)}(x)=Q_{\tau_0,\tau_1}^{(m)}\bar{v}_{\tau_1}^{(m)}(x)\vee f(\tau_0,x). \tag{26}$$

By using the fact that $|x\vee z-y\vee z|\leq|x-y|$ for all $x,y,z\in\mathbb{R}$, we have

$$\begin{aligned}|v_0(x)-\bar{v}_0^{(m)}(x)|&=|P_{\tau_0,\tau_1}v_{\tau_1}(x)\vee f(\tau_0,x)-Q_{\tau_0,\tau_1}^{(m)}\bar{v}_{\tau_1}^{(m)}(x)\vee f(\tau_0,x)|\\&\leq|P_{\tau_0,\tau_1}v_{\tau_1}(x)-Q_{\tau_0,\tau_1}^{(m)}\bar{v}_{\tau_1}^{(m)}(x)|\\&\leq|P_{\tau_0,\tau_1}(v_{\tau_1}(x)-\bar{v}_{\tau_1}^{(m)}(x))|+|(P_{\tau_0,\tau_1}-Q_{\tau_0,\tau_1}^{(m)})\bar{v}_{\tau_1}^{(m)}(x)|.\end{aligned} \tag{27}$$

By using Lemma 1 and the estimate $\|P_{\tau_0,\tau_1}g\|_\infty\leq\|g\|_\infty$ for a bounded measurable function g, the following inequality holds:

$$\|v_0-\bar{v}_0^{(m)}\|_\infty\leq\|v_{\tau_1}-\bar{v}_{\tau_1}^{(m)}\|_\infty+CK^{-(m+1)/2}\lambda^{m+1}\|\bar{v}_{\tau_1}^{(m)}\|_\infty. \tag{28}$$

Similarly, we get

$$\|v_{\tau_j} - \bar{v}^{(m)}_{\tau_j}\|_\infty \le \|v_{\tau_{j+1}} - \bar{v}^{(m)}_{\tau_{j+1}}\|_\infty + CK^{-(m+1)/2}\lambda^{m+1}\|\bar{v}^{(m)}_{\tau_{j+1}}\|_\infty, \quad (29)$$

for $j = 1, \dots K-1$. By using the fact that $v_{\tau_K} = \bar{v}^{(m)}_{\tau_K}$, we obtain

$$\|v_0 - \bar{v}^{(m)}_0\|_\infty \le C\sum_{j=1}^{K} K^{-(m+1)/2}\lambda^{m+1}\|\bar{v}^{(m)}_{\tau_j}\|_\infty. \quad (30)$$

Using the estimate $\|Q^{(m)}_{\tau_j,\tau_{j+1}}g\|_\infty \le \|g\|_\infty(1+c/K)$, for a bounded measurable function g and the fact $\bar{v}^{(m)}_{\tau_K} = f(\tau_K, \cdot)$, we have that there exists $C > 0$ which does not depend on K such that

$$\begin{aligned}
\|\bar{v}^{(m)}_{\tau_j}\|_\infty &= \|Q^{(m)}_{\tau_j,\tau_{j+1}}\bar{v}^{(m)}_{\tau_{j+1}} \vee f(\tau_j, \cdot)\|_\infty \\
&\le (1+c/K)\|\bar{v}^{(m)}_{\tau_{j+1}}\|_\infty \vee \|f(\tau_j, \cdot)\|_\infty \\
&\le \cdots \le (1+c/K)^K \max_{j\le i\le K}\|f(\tau_i, \cdot)\|_\infty \\
&\le e^c \max_{j\le i\le K}\|f(\tau_i, \cdot)\|_\infty \le C\max_{j\le i\le K}\|f(\tau_i, \cdot)\|_\infty, \quad (31)
\end{aligned}$$

for all $j = 1, \dots, K$. Finally, combining (30) with (31), it holds that for $m \in \mathbb{N}$,

$$\begin{aligned}
\|v_0 - \bar{v}^{(m)}_0\|_\infty &\le C\sum_{j=1}^{K} K^{-(m+1)/2}\lambda^{m+1}\max_{j\le i\le K}\|f(\tau_i, \cdot)\|_\infty \\
&\le C\max_{1\le j\le K}\|f(\tau_j, \cdot)\|_\infty\lambda^{m+1}. \ \square \quad (32)
\end{aligned}$$

References

1. Andersen, L., Broadie, M.: Primal-dual simulation algorithm for pricing multidimensional American options. Manage. Sci. **50**(9), 1222–1234 (2004)
2. Bally, V., Pagés, G., Printems, J.: A quantization tree method for pricing and hedging multidimensional American options. Math. Finance **15**(1), 119–168 (2005)
3. Beck, C., Becker, S., Cheridito, P., Jentzen, A., Neufeld, A: Deep splitting method for parabolic PDEs. SIAM J. Sci. Comput. **43**(5), 3135–3154 (2021)
4. Becker, S., Cheridito, P., Jentzen, A.: Deep optimal stopping. J. Mach. Learn. Res. **20**(1), 2712–2736 (2019)
5. Becker, S., Cheridito, P., Jentzen, A., Welti, T.: Solving high-dimensional optimal stopping problems using deep learning. Eur. J. Appl. Math. **32**(3), 470–514 (2021)
6. Belomestny, D.: Pricing Bermudan options by nonparametric regression: optimal rates of convergence for lower estimates. Finance Stoch. **15**(4), 655–683 (2011)

7. Clément, E., Lamberton, D., Protter, P.: An analysis of a least squares regression method for American option pricing. Finance Stoch. **6**(4), 449–471 (2002)
8. Fujii, M., Takahashi, A., Takahashi, M.: Asymptotic expansion as prior knowledge in deep learning method for high dimensional BSDEs. Asia-Pac. Financ. Markets **26**(3), 391–408 (2019)
9. Glasserman, P., Yu, B.: Number of paths versus number of basis functions in American option pricing. Ann. Appl. Probab. **14**(4), 2090–2119 (2004)
10. Han, J., Jentzen, A., Weinan E.: Solving high-dimensional partial differential equations using deep learning. Proc. Natl. Acad. Sci. **115**(34), 8505–8510 (2018)
11. Haugh, M.B., Kogan, L.: Pricing American options: a duality approach. Oper. Res. **52**(2), 258–270 (2004)
12. Huré, C., Pham, H., Warin, X.: Deep backward schemes for high-dimensional nonlinear PDEs. Math. Comput. **89**(324), 1547–1579 (2020)
13. Iguchi, Y., Yamada, T.: Operator splitting around Euler-Maruyama scheme and high order discretization of heat kernels. Math. Modell. Numer. Anal. **55**, S323–S367 (2021)
14. Ioffe, S., Szegedy, C.: Batch normalization: accelerating deep network training by reducing internal covariate shift. In: Proceedings of the 32nd International Conference on Machine Learning, vol. 37, pp. 448–456 (2015)
15. Kingma, D., Ba, J.: Adam: a method for stochastic optimization (2014). https://doi.org/10.48550/arXiv.1412.6980
16. Lapeyre, B., Lelong, J.: Neural network regression for Bermudan option pricing. Monte Carlo Methods Appl. **27**(3), 227–247 (2021)
17. Longstaff, F.A., Schwartz, E.S.: Valuing American options by simulation: a simple least-squares approach. Rev. Financ. Stud. **14**(1), 113–147 (2001)
18. Naito, R., Yamada, T.: A higher order weak approximation of McKean-Vlasov type SDEs. BIT Numer. Math. **62**, 521–559 (2022)
19. Naito, R., Yamada, T.: Pricing high-dimensional Bermudan options using deep learning and high-order weak approximation. J. Comput. Financ. **28**(1), 65–94 (2024)
20. Nualart, D.: The Malliavin Calculus and Related Topics. Springer, Berlin (2006)
21. Rogers, L.C.: Monte Carlo valuation of American options. Math. Financ. **12**(3), 271–286 (2002)
22. Sirignano, J., Spiliopoulos, K.: DGM: a deep learning algorithm for solving partial differential equations. J. Comput. Phys. **375**, 1339–1364 (2018)
23. Takahashi, A., Yamada, T.: An asymptotic expansion with push-down of Malliavin weights. SIAM J. Financ. Math. **3**, 95–136 (2012)
24. Takahashi, A., Yamada, T.: A weak approximation with asymptotic expansion and multidimensional Malliavin weights. Ann. Appl. Probab. **26**(2), 818–856 (2016)
25. Takahashi, A., Yamada, T.: New asymptotic expansion formula via Malliavin calculus and its application to rough differential equation driven by fractional Brownian motion. Asympt. Anal. **140**(1–2), 37–58 (2024)
26. Takahashi, A., Tsuchida, Y., Yamada, T.: A new efficient approximation scheme for solving high-dimensional semilinear PDEs: control variate method for Deep BSDE solver. J. Comput. Phys. **454**, 110956 (2022)
27. Tsitsiklis, J.N., Van Roy, B.: Regression methods for pricing complex American-style options. IEEE Trans. Neural Netw. **12**(4), 694–703 (2001)
28. Watanabe, S.: Analysis of Wiener functionals (Malliavin calculus) and its application to heat kernels. Ann. Probab. **15**(1), 1–39 (1987)
29. Yamada, T.: An arbitrary high order weak approximation of SDE and Malliavin Monte Carlo: analysis of probability distribution functions. SIAM J. Numer. Anal. **57**(2), 563–591 (2019)

Wong-Zakai Approximation for Stochastic PDEs and the HJM Model

Toshiyuki Nakayama

Abstract This lecture provides a clear overview of the collaborative paper with Professor Stefan Tappe titled "Wong-Zakai approximations with convergence rate for stochastic partial differential equations" (Nakayama and Tappe, Stoch. Anal. Appl. 36(5):832–857, 2018). It simplifies the core concepts and delves into my personal perspective on the significance of this paper in the field of computational finance.

Keywords Stochastic partial differential equation · Wong-Zakai approximation · Convergence rate · Graph norm

2010 Mathematics Subject Classification 60H15, 60G17

1 Introduction

We discuss the semi-linear stochastic partial differential equation (SPDE) of the form

$$\begin{cases} dX(t) = (AX(t) + b(X(t)))\,dt + \sum_{j=1}^{r} \sigma_j(X(t))dB^j(t), \\ X(0) = x_0 \end{cases} \tag{1.1}$$

on a separable Hilbert space $(H, \|\cdot\|)$, driven by a finite-dimensional Wiener process $B = (B^1, \ldots, B^r)$ for some positive integer r, where A is the infinitesimal generator of an arbitrary strongly continuous semigroup.

T. Nakayama (✉)
Shunan University, Yamaguchi, Japan
e-mail: nakayama_t25@shunan-u.ac.jp

J. Akahori et al. (eds.), *Recent Developments in Stochastic Numerics and Computational Finance*, ICIAM2023 Springer Series 6,
https://doi.org/10.1007/978-981-95-0652-1_4

A convergence result of the Wong-Zakai approximation has first been proven in the case of finite-dimensional SDEs by Wong and Zakai [30]. The Wong-Zakai approximation represents an equation obtained by replacing the Wiener process with an appropriate function that has a well-defined time derivative, and shifting the drift term. The Wong-Zakai approximation is useful for constructing numerical calculation techniques for stochastic differential equations ([7, 18, 19] for example).

The Wong-Zakai approximation has been generalized into the infinite-dimensional case by many authors ([1, 2, 5, 6, 12–16, 20, 24] and [25–29] for example). Regarding convergence rates, there are some results such as for a second-order parabolic type [13, 14] and for the infinitesimal generator of a compact and analytic semigroup [16]. In this lecture, we explain a more general result [22] in the sense that A is allowed to be the infinitesimal generator of an arbitrary strongly continuous semigroup.

Finally, we will introduce an application example of SPDEs in the HJM (Heath-Jarrow-Morton) model [17], which is an important interest rate model in mathematical finance. The HJM model describes the price $P(t, T)$ of the zero-coupon bond with maturity T observed at time t by

$$P(t, T) = \exp\left(- \int_t^T f(t, u) du \right),$$

where $f(t, T)$ is the instantaneous forward rate at date T observed at time t given by the following stochastic differential equation (SDE)

$$df(t, T) = \sum_{j=1}^{n} \sigma_j(t, T) \int_t^T \sigma_j(t, u) du dt + \sum_{j=1}^{n} \sigma_j(t, T) dB^j(t), \tag{1.2}$$

where $(B(t))_{t\in[0,T]}$ $(B(t) = (B^1(t), \ldots, B^n(t)))$ is an $\mathbb{R}^n$-valued Wiener process for an integer n with respect to a risk-neutral measure, and $(\sigma_j(t, T))_{t\in[0,T]}$ is a real-valued stochastic process for each $T \geq 0$ and $j = 1, \ldots, n$ that satisfies some appropriate conditions. The term $\sigma_j(t, T)$ is called volatility for each $j = 1, \ldots, n$. The drift term in the SDE (1.2) comes from the condition that there is no arbitrage opportunity with a portfolio consisting of the money market account $\exp(\int_0^t f(u, u) du)$ and the zero-coupon bonds of various maturities.

There are many kinds of models corresponding to the choice of volatilities $\sigma_j(t, T)$ with $j = 1, \ldots, n$. A simple example for $n = 1$ is

$$\sigma(t, T) = g(f(t, t), f(t, T)), \quad 0 \leq t \leq T$$

for a real-valued function g. As this example illustrates, for each $t \geq 0$, the volatility may generally depend on the function $u \mapsto f(t, u)$ with $u \in [t, \infty)$. Since the domain $[t, \infty)$ of this function depends on t, which we want to avoid for simplicity, let us change the variable by

$$(t, T) \mapsto (t, x) = (t, T - t), \quad t \geq 0, \; T \geq t$$

and introduce the instantaneous forward rate with maturity time x from observing time t as

$$r(t)(x) := r(t, x) := f(t, t + x), \quad t \geq 0, \; x \geq 0.$$

This is called Musiela parametrization (see [4]) and, for each $t \geq 0$, the function $r(t)\colon x \mapsto r(t)(x)$ with $x \in [0, \infty)$ is called 'forward curve'.

The volatility $\sigma(t, T)$ in the above example is expressed as

$$\sigma(t, t + x) = g(r(t)(0), r(t)(x)) = \gamma(r(t))(x),$$

where

$$\gamma(h)(x) := g(h(0), h(x)).$$

The term 'volatility' mentioned above is also used for $\gamma(r(t))(x)$. Similar to the forward curve, let us call the function $\gamma(r(t))\colon x \mapsto \gamma(r(t))(x)$ with $x \in [0, \infty)$ 'volatility curve'.

After transforming the original HJM dynamics (1.2) by means of Musiela parametrization, we get the following SPDE called HJMM (Heath-Jarrow-Morton-Musiela) equation

$$\begin{cases} dr(t) = \dfrac{\partial}{\partial x} r(t)dt + \alpha(r(t))dt + \displaystyle\sum_{j=1}^{n} \gamma_j(r(t))dB^j(t), \\ r(0) = r_0, \end{cases} \tag{1.3}$$

where

$$\alpha(h) = \sum_{j=1}^{n} \gamma_j(h) \int_0^{\bullet} \gamma_j(h)(\eta)d\eta. \tag{1.4}$$

The SPDE (1.3) can be formulated as an infinite-dimensional SDE, where the state space is an infinite-dimensional separable Hilbert space H consisting of forward curves.

Future forward curves, which are generated by the infinite-dimensional SDE, are often modeled to remain within a finite-dimensional submanifold of H, which offers advantages in terms of efficiency for numerical computations (see, for example, [21] for the definition of finite-dimensional submanifold). In particular, there are cases where models are preferred to ensure that future forward curves remain within a finite dimensional submanifold $\mathcal{M}$ comprised of the initial forward curves that are conceivable candidates for r_0 in the SPDE (1.3). Here the finite dimensional submanifold $\mathcal{M}$ is determined by how initial forward curves are constructed, with parametric families like the Nelson-Siegel family and the Svensson family being well-known examples. The condition of remaining within $\mathcal{M}$ is expressed by the following equation

$$\mathbb{P}\left(\exists t \geq 0,\ r^{(r_0)}(t) \notin \mathcal{M}\right) = 0 \text{ for all } r_0 \in \mathcal{M},$$

where $r^{(r_0)}$ is, in a sense, a solution to the SPDE (1.3) with $r(0) = r_0$. Such models are said to satisfy 'consistency' [3, 10, 11, 21] and align with the practical practice of updating the initial curves every business day.

However, in general, the forward curves from the SPDE (1.3) may not necessarily remain within a finite-dimensional submanifold. It is common practice to estimate the coefficients γ_j with $j = 1, \ldots, n$ of the SPDE (1.3) so that they fit well with the prices of financial products observed for many maturities in the market. As a result, the forward curve, whether we like it or not, might end up being modeled as essentially distributed in an infinite-dimensional space in the sense that it does not fit into any finite-dimensional submanifold. Our theorem of convergence order, of course, holds true for this infinite-dimensional case.

On the other hand, note that the Wiener process in SPDE (1.3) is finite-dimensional, which is a reasonable assumption from a practical point of view. But some problems arise from a theoretical point of view. Let us consider an example of replicating a payoff ξ at time $T > 0$ using the money market account and the zero-coupon bonds. Let us select n dates $T_1 < T_2 < \cdots < T_n$ with $T_1 > T$. If the matrix

$$\left(P(t, T_i) \int_0^{T_i - t} \gamma_j(r(t))(\eta) d\eta\right)_{1 \leq i, j \leq n},$$

which comes from the volatilities of the zero-coupon bond prices, is invertible for almost all $(t, \omega) \in [0, T] \times \Omega$, then the payoff ξ satisfying an appropriate integrability condition can be replicated by a self-financing portfolio consisting of the money market account $\exp\left(\int_0^t r(s)(0) ds\right)$ and the zero-coupon bonds $P(t, T_1)$, $P(t, T_2), \ldots, P(t, T_n)$ (see [8, 2.4.2] for details). From this, $P(t, T)$ can be replicated not only in a trivial sense with $P(t, T)$, but also with the money market account and $P(t, T_1)$, $P(t, T_2), \ldots, P(t, T_n)$, hence the replicating portfolio is not unique. For example, if interest rates are modeled using a 2-dimensional Wiener process, which is often seen in practice, the zero-coupon bond with the maturity of 1 year might be replicated using a combination of the money market account, the zero-coupon bond with the maturity of 10 years and the zero-coupon bond with the maturity of 20 years. This might seem counterintuitive to traders.

An effective method to avoid such inconvenience is to introduce an infinite-dimensional Wiener process. Typically, we use a U-valued Q-Wiener process, where U is a separable Hilbert space, and Q is a nuclear, self-adjoint, positive definite linear operator. Volatilities are then replaced by a Hilbert-Schmidt operator from $Q^{1/2}(U)$ to H. A stochastic integral can be defined within this framework (see [9] for example). The use of this infinite-dimensional Wiener process is highly effective in theoretical considerations of interest rate models.

However, from the perspective of numerical calculation efficiency, many practitioners prefer low-dimensional Wiener processes, typically, fewer than ten for market risk management, and even two or three in pricing, etc. This preference is supported by principal component analysis and the idea that one of the roles of many mathematical models is to simply capture the real world. We believe that our result for finite-dimensional Wiener process is meaningful in that it supports the development of computational calculation technology for models used in practice.

The remainder of this lecture is organized as follows. In Sect. 2, we state a theorem (Theorem 2.6) concerning a convergence rate of the Wong-Zakai approximation of the SPDE, which is our main topic. We claim that the proof of this theorem reduces to the proof of two other theorems (Theorems 2.7 and 2.8). In Sect. 3, we provide a sketch of the proof of Theorem 2.7. In Sect. 4, we provide a sketch of the proof of Theorem 2.8. In Sect. 5, we provide an example of an application to the SPDE regarding to the HJM model in a more general setting stated above, which includes the local volatility model and the stochastic volatility model.

2 Convergence Order of Wong-Zakai Approximation for SPDE

2.1 General Framework

First, we prepare for the formulation of the SPDE.

Let be a C_0-semigroup on a separable Hilbert space H with the infinitesimal generator $A\colon \mathcal{D}(A) \subset H \to H$.

Remark 2.1

- The space $(\mathcal{D}(A), \|\cdot\|_{\mathcal{D}(A)})$ is a separable Hilbert space, where $\|\cdot\|_{\mathcal{D}(A)}$ is the graph norm, i.e. $\|x\|_{\mathcal{D}(A)} = \sqrt{\|x\|^2 + \|Ax\|^2}$ for $x \in \mathcal{D}(A)$.
- The restriction $(S_t|_{\mathcal{D}(A)})_{t\in[0,T]}$ is a C_0-semigroup on $(\mathcal{D}(A), \|\cdot\|_{\mathcal{D}(A)})$ with the infinitesimal generator $A|_{\mathcal{D}(A^2)}$.

Let us consider the following SPDE on H

$$\begin{cases} dX(t) = (AX(t) + b(X(t)))\,dt + \sum_{j=1}^{r} \sigma_j(X(t))dB^j(t), \\ X(0) = x_0. \end{cases} \tag{2.1}$$

Here $x_0 \in \mathcal{D}(A)$ is a constant vector, and $(B^j(t))_{t\geq 0}$, $j = 1, \ldots, r$ are independent real-valued standard Wiener processes on a probability space $(\Omega, \mathcal{F}, (\mathcal{F})_{t\in[0,T]}, \mathbb{P})$

with a filtration satisfying the usual conditions. We make assumptions on the coefficients as follows.

Assumption 2.2 We suppose that the following conditions are fulfilled:

- The drift $b\colon H \to H$ is Lipschitz continuous and bounded.
- We have $\sigma_j \in C_b^2(H)$, i.e. twice continuously Fréchet differentiable as a mapping from H to H, and both σ_j, $D\sigma_j$ and $D^2\sigma_j$ are bounded for each $j = 1, \ldots, r$.

Under this assumption, the SPDE (2.1) has a unique mild solution on H (see [9, Theorem 7.2] for example), where H-valued predictable process $(X(t))_{t\geq 0}$ is a mild solution if

$$X(t) = S_t x_0 + \int_0^t S_{t-s} b(X(s))ds + \sum_{j=1}^r \int_0^t S_{t-s}\sigma_j(X(s))dB^j(s)$$

for $t \in [0, T]$ almost surely. However, in general, there does not exist a strong solution on H, where H-valued predictable process $(X(t))_{t\geq 0}$ is a strong solution if $(X(t))_{t\geq 0}$ satisfies $X(t) \in \mathcal{D}(A)$ almost surely for $t \in [0, T]$ and

$$X(t) = x_0 + \int_0^t (AX(s) + b(X(s)))\,ds + \sum_{j=1}^r \int_0^t \sigma_j(X(s))dB^j(s)$$

almost surely for $t \in [0, T]$.

Therefore, let us assume additionally the following.

Assumption 2.3 We suppose that the following conditions are fulfilled:

- We have $b(\mathcal{D}(A)) \subset \mathcal{D}(A)$ and $\sigma_j(\mathcal{D}(A)) \subset \mathcal{D}(A)$ for $j = 1, \ldots, r$.
- The restriction $b|_{\mathcal{D}(A)}\colon \mathcal{D}(A) \to \mathcal{D}(A)$ is Lipschitz continuous and bounded with respect to the graph norm $\|\cdot\|_{\mathcal{D}(A)}$.
- The restriction satisfies $\sigma_j|_{\mathcal{D}(A)} \in C_b^2(\mathcal{D}(A))$ for each $j = 1, \ldots, r$ with respect to the graph norm $\|\cdot\|_{\mathcal{D}(A)}$.

By making this assumption, the following proposition can be obtained.

Proposition 2.4 *For each $x_0 \in \mathcal{D}(A)$, there exists a unique mild solution $(X(t))_{t\in[0,T]}$ to the SPDE (2.1) on the state space $(\mathcal{D}(A), \|\cdot\|_{\mathcal{D}(A)})$, and it is a strong solution to the SPDE (2.1) on the state space $(H, \|\cdot\|)$.*

Proof See [22, Proposition 2.8]. □

2.2 Wong-Zakai Approximation

Let $T > 0$ be fixed. For each $m \in \mathbb{N}$, let us define

$$\delta_m := \frac{T}{m}, \quad m \in \mathbb{N},$$

$$[t]_m^- := k\delta_m, \quad k\delta_m \le t < (k+1)\delta_m, \quad k = 0, \dots, m-1,$$

$$[t]_m^+ := (k+1)\delta_m, \quad k\delta_m \le t < (k+1)\delta_m, \quad k = 0, \dots, m-1,$$

$$B_m^j(t) := B^j([t]_m^-) + \frac{t - [t]_m^-}{\delta_m}(B^j([t]_m^+) - B^j([t]_m^-)), \quad t \in [0, T].$$

The stochastic process $(B_m^j(t))_{t\in[0,T]}$ is interpreted as a linear interpolation of the Wiener process $(B^j(t))_{t\in[0,T]}$.

The Wong-Zakai approximation is defined by the following PDE

$$\begin{cases} \dot{\xi}_m(t) = A\xi_m(t) + b(\xi_m(t)) - \rho(\xi_m(t)) + \sum_{j=1}^{r} \sigma_j(\xi_m(t))\dot{B}_m^j(t), \\ \xi_m(0) = x_0 \end{cases} \tag{2.2}$$

for each $m \in \mathbb{N}$, where from Assumption 2.2, the mapping $\rho\colon H \to H$ defined by

$$\rho(x) = \frac{1}{2}\sum_{j=1}^{r} D\sigma_j(x)\sigma_j(x)$$

is Lipschitz continuous and bounded. From Assumption 2.3, the restriction $\rho|_{\mathcal{D}(A)}\colon \mathcal{D}(A) \to \mathcal{D}(A)$ is also Lipschitz continuous and bounded with respect to the graph norm $\|\cdot\|_{\mathcal{D}(A)}$.

Proposition 2.5 *For each $x_0 \in \mathcal{D}(A)$ and each $m \in \mathbb{N}$, there exists a unique mild solution $(\xi_m(t))_{t\in[0,T]}$ to the PDE (2.2) on the state space $(\mathcal{D}(A), \|\cdot\|_{\mathcal{D}(A)})$, and it is a strong solution to the PDE (2.2) on the state space $(H, \|\cdot\|)$.*

Proof See [22, Proposition 2.9]. □

2.3 Main Theorem

Let us fix $p > 1$. It has been proven that the Wong-Zakai approximation $(\xi_m)_{m\in\mathbb{N}}$ converges to the solution X of the SPDE (2.1) [20, Theorem 2.1] in the sense that

$$\lim_{m\to\infty} \mathbb{E}\left[\sup_{t\in[0,T]} \|\xi_m(t) - X(t)\|^{2p}\right] = 0, \tag{2.3}$$

where Assumption 2.3 is not required. The main purpose of this lecture is to explain the key points of the proof of the following theorem, which claims the convergence rate of the Wong-Zakai approximation.

Theorem 2.6 *Suppose that Assumptions 2.2 and 2.3 are fulfilled, and let $T > 0$, $p > 1$ and $x_0 \in \mathcal{D}(A)$ be arbitrary. Then there is a constant $C > 0$ such that for each $m \in \mathbb{N}$ we have*

$$\mathbb{E}\left[\sup_{t\in[0,T]} \|\xi_m(t) - X(t)\|^{2p}\right] \le \frac{C}{m^{p-1}}, \tag{2.4}$$

where X denotes the mild solution to the SPDE (2.1) with $X(0) = x_0$, and $(\xi_m)_{m\in\mathbb{N}}$ denote the mild solutions to the PDE (2.2) with $\xi_m(0) = x_0$.

2.4 Outline of the Proof of Theorem 2.6

By replacing symbols b and $b - \rho$ in Eqs. (2.1) and (2.2) defining the above X and ξ_m anew with $\hat{b} = b + \rho$ and b, respectively, we can consider the following case without loss of generality.

$$\begin{cases} dX(t) = \left(AX(t) + \hat{b}(X(t))\right) dt + \displaystyle\sum_{j=1}^{r} \sigma_j(X(t))dB^j(t), \\ X(0) = x_0, \end{cases} \tag{2.5}$$

and

$$\begin{cases} \dot{\xi}_m(t) = (A\xi_m(t) + b(\xi_m(t)))\, dt + \displaystyle\sum_{j=1}^{r} \sigma_j(\xi_m(t))\dot{B}_m^j(t), \\ \xi_m(0) = x_0. \end{cases} \tag{2.6}$$

Let $(Y_m(t))_{t\in[0,T]}$ be the stochastic process, called Euler-Muruyama approximation, defined by

$$\begin{cases} Y_m(t) = S_{t-k\delta_m} Y_m(k\delta_m) + \displaystyle\int_{k\delta_m}^{t} S_{t-s}\hat{b}(Y_m(k\delta_m))ds \\ \quad + \displaystyle\sum_{j=1}^{r} \int_{k\delta_m}^{t} S_{t-s}\sigma_j(Y_m(k\delta_m))dB^j(s), \quad t \in [k\delta_m, (k+1)\delta_m], \\ Y_m(0) = x_0 \end{cases} \tag{2.7}$$

for $k = 0, 1, \ldots, m-1$. Then we have

$$\begin{aligned} Y_m(t) = S_t x_0 &+ \int_0^t S_{t-s}\hat{b}(Y_m([s]_m^-))ds \\ &+ \sum_{j=1}^r \int_0^t S_{t-s}\sigma_j(Y_m([s]_m^-))dB^j(s), \quad t \in [0,T]. \end{aligned} \tag{2.8}$$

Our goal of the proof of Theorem 2.6 is achieved by proving two key theorems. First, we will obtain the convergence rate of the Euler-Maruyama approximation.

Theorem 2.7 *Suppose that Assumptions 2.2 and 2.3 are fulfilled, and let* $T > 0$, $p > 1$ *and* $x_0 \in \mathcal{D}(A)$ *be arbitrary. Then there is a constant* $C > 0$ *such that for each* $m \in \mathbb{N}$ *we have*

$$\mathbb{E}\left[\sup_{t\in[0,T]} \|Y_m(t) - X(t)\|^{2p}\right] \leq \frac{C}{m^{p-1}}, \tag{2.9}$$

where X *denotes the mild solution to the SPDE (2.5) with* $X(0) = x_0$, *and* $(Y_m)_{m\in\mathbb{N}}$ *denote the Euler-Maruyama approximations given by (2.7) with* $Y_m(0) = x_0$.

Furthermore, we will obtain the convergence rate of the distance of the Euler-Maruyama and Wong-Zakai approximations.

Theorem 2.8 *Suppose that Assumptions 2.2 and 2.3 are fulfilled, and let* $T > 0$, $p > 1$ *and* $x_0 \in \mathcal{D}(A)$ *be arbitrary. Then there is a constant* $C > 0$ *such that for each* $m \in \mathbb{N}$ *we have*

$$\mathbb{E}\left[\sup_{t\in[0,T]} \|\xi_m(t) - Y_m(t)\|^{2p}\right] \leq \frac{C}{m^{p-1}}, \tag{2.10}$$

where $(\xi_m)_{m\in\mathbb{N}}$ *denote the mild solutions to the PDE (2.6) with* $\xi_m(0) = x_0$, *and* $(Y_m)_{m\in\mathbb{N}}$ *denote the Euler-Maruyama approximations given by (2.7) with* $Y_m(0) = x_0$.

3 Sketch of the Proof of Theorem 2.7

Let us fix arbitrary $T > 0$, $p > 1$ and $x_0 \in \mathcal{D}(A)$. In the following discussion, we will denote by C an appropriate positive constant, possibly different from line to line.

To prove Theorem 2.7, let us first prepare lemmas.

Note that it is well-known that there are constants $M \geq 1$ and $\omega \in \mathbb{R}$ such that

$$\|S_t\| \le Me^{\omega t} \quad \text{for all } t \ge 0 \tag{3.1}$$

(see [23, Thm. 2.2] for example).

Lemma 3.1 *Let* $j \in \{1, \dots, r\}$*. For any* H*-valued predictable process* $(\Phi(t))_{t\in[0,T]}$ *such that*

$$\mathbb{E}\left[\int_0^T \|\Phi(s)\|^{2p}\, ds\right] < \infty,$$

there exists $C_p > 0$ *such that*

$$\mathbb{E}\left[\sup_{0\le t\le T}\left\|\int_0^t \Phi(s)dB^j(s)\right\|^{2p}\right] \le C_p\mathbb{E}\left[\left(\int_0^T \|\Phi(s)\|^2\, ds\right)^p\right].$$

Proof See Theorem 4.36 in Da Prato and Zabczyk [9] and apply Jensen's inequality. □

The following lemma comes from [20, Lemma 2.4].

Lemma 3.2 *Let* $j \in \{1, \dots, r\}$*. For any* H*-valued predictable process* $(\gamma(t))_{t\in[0,T]}$ *where* $\sup\{\|\gamma(t,\omega)\|;\ 0 \le t \le T,\ \omega \in \Omega\} < \infty$*, there exists a constant* $C > 0$ *such that*

$$E\left[\sup_{0\le t\le T} \| \int_{[t]_m^-}^t S(t-s)\gamma(s)dB^j(s)\|^{2p}\right] \le C\delta_m^{p-1}.$$

Proof Note that the integrand in the stochastic integral $\int_{[t]_m^-}^t S(t-s)\gamma(s)dB^j(s)$ depends on t. This means that inequalities like those in Lemma 3.1 cannot be directly used for evaluation, and special considerations are required. Therefore, to conduct a discussion similar to Chapter 5 in Da Prato and Zabczyk [9], let us fix a real number $\alpha \in \left(\frac{1}{2p}, \frac{1}{2}\right)$ and note the identity

$$\int_s^t (t-u)^{\alpha-1}(u-s)^{-\alpha}du = \frac{\pi}{\sin \pi\alpha}, \qquad 0 \le s < t \le T. \tag{3.2}$$

By this identity (3.2) and the stochastic Fubini theorem, we have

$$\int_{[t]_m^-}^t S_{t-s}\gamma(s)dB^j(s)$$
$$= \frac{\sin \pi\alpha}{\pi}\int_{[t]_m^-}^t (t-u)^{\alpha-1}S_{t-u}\Big(\int_{[t]_m^-}^u (u-s)^{-\alpha}S_{u-s}\gamma(s)dB^j(s)\Big)du.$$

So, we have

$$\| \int_{[t]_m^-}^t S_{t-s}\gamma(s)dB^j(s)\| \le \frac{\sin \pi\alpha}{\pi} \sup_{0\le u\le T} \|S_u\| \int_{[t]_m^-}^t (t-u)^{\alpha-1} \| \int_{[t]_m^-}^u (u-s)^{-\alpha} S_{u-s}\gamma(s)dB^j(s)\|du.$$

By the inequality (3.1) and Hölder inequality, we have

$$\begin{aligned}
&\| \int_{[t]_m^-}^t S_{t-s}\gamma(s)dB^j(s)\|^{2p} \\
&\le C\delta_m^{2p\alpha-1} \int_{[t]_m^-}^t \| \int_{[t]_m^-}^u (u-s)^{-\alpha} S_{u-s}\gamma(s)dB^j(s)\|^{2p}du \\
&\le C\delta_m^{2p\alpha-1} \sum_{k=0}^{m-1} \int_{k\delta_m}^{(k+1)\delta_m} \| \int_{k\delta_m}^u (u-s)^{-\alpha} S_{u-s}\gamma(s)dB^j(s)\|^{2p}du
\end{aligned}$$

for all $t \in [0, T]$. Therefore, by using Lemma 3.1, we get

$$\begin{aligned}
&E\Big[\sup_{0\le t\le T} \| \int_{[t]_m^-}^t S_{t-s}\gamma(s)dB^j(s)\|^{2p}\Big] \\
&\le C\delta_m^{2p\alpha-1} \sum_{k=0}^{m-1} \int_{k\delta_m}^{(k+1)\delta_m} E\Big[\Big(\int_{k\delta_m}^u (u-s)^{-2\alpha} \|S_{u-s}\gamma(s)\|^2 ds\Big)^p\Big]du \\
&\le C\delta_m^{2p\alpha-1} \sum_{k=0}^{m-1} \int_{k\delta_m}^{(k+1)\delta_m} \Big(\frac{\delta_m^{1-2\alpha}}{1-2\alpha}\Big)^p du \le C\delta_m^{p-1}.
\end{aligned}$$

□

Now let us estimate $Y_m(t) - X(t)$ to prove Theorem 2.7. Let us define the process $(\bar{Y}_m(t))_{t\ge0}$ and $(\bar{X}_m(t))_{t\ge0}$ as

$$\begin{aligned}
\bar{Y}_m(t) &:= S_{t-[t]_m^-} Y_m([t]_m^-), \quad t \ge 0, \\
\bar{X}_m(t) &:= S_{t-[t]_m^-} X([t]_m^-), \quad t \ge 0.
\end{aligned}$$

Then we have the decomposition

$$Y_m - X = (Y_m - \bar{Y}_m) + (\bar{Y}_m - \bar{X}_m) + (\bar{X}_m - X). \tag{3.3}$$

The first term $Y_m - \bar{Y}_m$ in the right-hand side can be estimated as follows.

Proposition 3.3 *There is a constant $C > 0$ such that for each $m \in \mathbb{N}$ we have*

$$\mathbb{E}\left[\sup_{t\in[0,T]} \|Y_m(t) - \bar{Y}_m(t)\|^{2p}\right] \leq C\delta_m^{p-1}.$$

Proof We have

$$Y_m(t) - \bar{Y}_m(t) = \int_{[t]_m^-}^{t} S_{t-s}\hat{b}(Y_m([s]_m^-))ds + \sum_{j=1}^{r}\int_{[t]_m^-}^{t} S_{t-s}\sigma_j(Y_m([s]_m^-))dB^j(s).$$

Since

$$\left\|\int_{[t]_m^-}^{t} S_{t-s}\hat{b}(Y_m([s]_m^-))ds\right\| \leq C\delta_m,$$

applying Lemma 3.2 completes the proof. □

For the term $\bar{X}_m - X$ in the Eq. (3.3), we have the following estimation.

Proposition 3.4 *There is a constant $C > 0$ such that for each $m \in \mathbb{N}$ we have*

$$\mathbb{E}\left[\sup_{t\in[0,T]} \|\bar{X}_m(t) - X(t)\|^{2p}\right] \leq C\delta_m^{p-1}.$$

Proof We can use a similar argument to Proposition 3.3 using Lemma 3.2. See [20, Lemma 2.5] for details. □

For the term $\bar{Y}_m - \bar{X}_m$ in the Eq. (3.3), we have

Proposition 3.5 *There is a constant $C > 0$ such that for each $m \in \mathbb{N}$ and each $v \in [0, T]$ we have*

$$\mathbb{E}\left[\sup_{t\in[0,v]} \|\bar{Y}_m(t) - \bar{X}_m(t)\|^{2p}\right] \leq C\left(\int_0^v \mathbb{E}\left[\sup_{t\in[0,u]} \|Y_m(t) - X(t)\|^{2p}\right]du + \delta_m^{p-1}\right).$$

Proof See [22, Proposition 3.5]. □

From Propositions 3.3, 3.4, 3.5 and the Eq. (3.3), Theorem 2.7 can be obtained using Gronwall's inequality.

4 Sketch of the Proof of Theorem 2.8

As in the previous sections, we will focus on presenting the characteristic proofs to prove Theorem 2.8 in an easy-to-understand manner, so please refer to the original paper [22] for the remaining details.

Let us fix arbitrary $T > 0$, $p > 1$ and $x_0 \in \mathcal{D}(A)$. In the following discussion, we will denote by C an appropriate positive constant, possibly different from line to line.

Let us fix $m \in \mathbb{N}$ and define processes $(\bar{\xi}_m(t))_{t\geq 0}$ and $(\bar{Y}_m(t))_{t\geq 0}$ by

$$\bar{\xi}_m(t) := S_{t-[t]_m^-}\xi_m([t]_m^-), \quad t \geq 0,$$

$$\bar{Y}_m(t) := S_{t-[t]_m^-}Y_m([t]_m^-), \quad t \geq 0.$$

Then, for each $m \in \mathbb{N}$, we have the decomposition

$$\xi_m - Y_m = (\xi_m - \bar{\xi}_m) + (\bar{\xi}_m - \bar{Y}_m) + (\bar{Y}_m - Y_m). \tag{4.1}$$

For the term $\bar{Y}_m - Y_m$, we have already an estimation of Proposition 3.3. And for the term $\xi_m - \bar{\xi}$, we also have the following estimation.

Proposition 4.1 *There is a constant $C > 0$ such that for each $m \in \mathbb{N}$ we have*

$$\mathbb{E}\left[\sup_{t\in[0,T]} \|\xi_m - \bar{\xi}_m\|^{2p}\right] \leq C\delta_m^{p-1}.$$

Proof See [20, Lemma 2.2]. □

In the following, we will concentrate on the term $\bar{\xi}_m - \bar{Y}_m$ in the Eq. (4.1). Because of the identity

$$\bar{\xi}_m(t) - \bar{Y}_m(t) = S_{t-[t]_m^-}\big(\xi_m([t]_m^-) - Y_m([t]_m^-)\big),$$

we have

$$\begin{aligned}
\bar{\xi}_m(t) - \bar{Y}_m(t) = S_{t-[t]_m^-}\Bigg(&\int_0^{[t]_m^-} S_{t-s}b(\xi_m(s))ds \\
&+\sum_{j=1}^{r}\int_0^{[t]_m^-} S_{t-s}\sigma_j(\xi_m(s))\dot{B}_m^j(s)ds - \int_0^{[t]_m^-} S_{t-s}b(Y_m([s]_m^-))ds \\
&-\frac{1}{2}\sum_{j=1}^{r}\int_0^{[t]_m^-} S_{t-s}D\sigma_j(Y_m([s]_m^-))\sigma_j(Y_m([s]_m^-))ds \\
&-\sum_{j=1}^{r}\int_0^{[t]_m^-} S_{t-s}\sigma_j(Y_m([s]_m^-))dB^j(s)\Bigg).
\end{aligned} \tag{4.2}$$

Let us fix an arbitrary $j \in \{1, \ldots, r\}$. In the right-hand side of this equation, noting the identity

$$\sigma_j(h) = \sigma_j(h_0) + D\sigma_j(h_0)(h - h_0) + \int_0^1 \left(\int_0^{v_1} D^2\sigma_j(h_0 + v_2(h - h_0))(h - h_0, h - h_0)dv_2 \right) dv_1$$

for all $h, h_0 \in H$, we can express the term $\sigma_j(\xi_m(t))$ as

$$\sigma_j(\xi_m(t)) = \gamma_{m,1}^j(t) + \sum_{l=1}^r \gamma_{m,2}^{j,l}(t) + \gamma_{m,3}^j(t), \quad t \in [0, T]. \tag{4.3}$$

Here the quantity $\gamma_{m,1}^j(t)$ is given by

$$\gamma_{m,1}^j(t) := \sigma_j(\bar{\xi}_m(t)),$$

the quantities $\gamma_{m,2}^{j,l}(t)$ for $l = 1, \ldots, r$ are given by

$$\gamma_{m,2}^{j,l}(t) := D\sigma_j(\bar{\xi}_m(t)) \int_{[t]_m^-}^t S_{t-u}\sigma_l(\bar{\xi}_m(u))\dot{B}_m^l(u)du,$$

and the quantity $\gamma_{m,3}^j(t)$ is given by

$$\begin{aligned}
\gamma_{m,3}^j(t) := {} & D\sigma_j(\bar{\xi}_m(t)) \int_{[t]_m^-}^t S_{t-u}b(\xi_m(u))du \\
& + \sum_{l=1}^r D\sigma_j(\bar{\xi}_m(t)) \int_{[t]_m^-}^t S_{t-u}\Bigg(\int_0^1 D\sigma_l(\bar{\xi}_m(u) \\
& + v(\xi_m(u) - \bar{\xi}_m(u)))(\xi_m(u) - \bar{\xi}_m(u))dv \Bigg) \dot{B}_m^l(u)du \\
& + \int_0^1 \Bigg(\int_0^{v_1} D^2\sigma_j(\bar{\xi}_m(t) + v_2(\xi_m(t) - \bar{\xi}_m(t))) \\
& (\xi_m(t) - \bar{\xi}_m(t), \xi_m(t) - \bar{\xi}_m(t)))dv_2 \Bigg) dv_1.
\end{aligned}$$

By substituting the expansion (4.3) into the right-hand side of the Eq. (4.2), we get

$$\bar{\xi}_m(t) - \bar{Y}_m(t) = S_{t-[t]_m^-}$$
$$\Big(\sum_{j=1}^{r} \zeta_{m,1}^{j}(t) + \sum_{j=1}^{r} \zeta_{m,2}^{j}(t) + \sum_{j=1}^{r}\sum_{l=1}^{r} \zeta_{m,3}^{j,l}(t) + \sum_{j=1}^{r} \zeta_{m,4}^{j}(t) + \zeta_{m,5}(t) \Big). \tag{4.4}$$

Here we introduced the quantity $\zeta_{m,1}^{j}(t)$ as

$$\zeta_{m,1}^{j}(t) := \int_0^{[t]_m^-} S_{t-s}\gamma_{m,3}^{j}(s)\dot{B}_m^{j}(s)ds,$$

the quantity $\zeta_{m,2}^{j}(t)$ as

$$\zeta_{m,2}^{j}(t) := \int_0^{[t]_m^-} S_{t-s}\gamma_{m,1}^{j}(s)\dot{B}_m^{j}(s)ds - \int_0^{[t]_m^-} S_{t-s}\sigma_j(Y_m([s]_m^-))dB^j(s),$$

the quantities $\zeta_{m,3}^{j,l}(t)$ for $l = 1, \dots, r$ as

$$\zeta_{m,3}^{j,j}(t) := \int_0^{[t]_m^-} S_{t-s}\gamma_{m,2}^{j,j}(s)\dot{B}_m^{j}(s)ds$$
$$- \frac{1}{\delta_m}\int_0^{[t]_m^-} S_{t-s}D\sigma_j(\bar{\xi}_m(s))\Big(\int_{[s]_m^-}^{s} S_{s-u}\sigma_j(\bar{\xi}_m(u))du\Big)ds$$

and

$$\zeta_{m,3}^{j,l}(t) := \int_0^{[t]_m^-} S_{t-s}\gamma_{m,2}^{j,l}(s)\dot{B}_m^{j}(s)ds \quad \text{for } l \neq j,$$

the quantity $\zeta_{m,4}^{j}(t)$ as

$$\zeta_{m,4}^{j}(t) := \frac{1}{\delta_m}\int_0^{[t]_m^-} S_{t-s}D\sigma_j(\bar{\xi}_m(s))\Big(\int_{[s]_m^-}^{s} S_{s-u}\sigma_j(\bar{\xi}_m(u))du\Big)ds$$
$$- \frac{1}{2}\int_0^{[t]_m^-} S_{t-s}D\sigma_j(Y_m([s]_m^-))\sigma_j(Y_m([s]_m^-))ds,$$

and the quantity $\zeta_{m,5}(t)$ as

$$\zeta_{m,5}(t) := \int_0^{[t]_m^-} S_{t-s}\big(b(\xi_m(s)) - b(Y_m([s]_m^-))\big)ds.$$

We have the following estimation for all terms except $\zeta_{m,3}^{j,l}(t)$ for $l \in \{1, \dots, r\}$.

Proposition 4.2 *There is a constant $C > 0$ such that for each $m \in \mathbb{N}$ we have*

$$\mathbb{E}\Big[\sup_{t\in[0,T]} \|\zeta^j_{m,1}(t)\|^{2p}\Big] \le C\delta^p_m,$$

and for each $v \in [0, T]$ we also have

$$\mathbb{E}\Big[\sup_{t\in[0,v]} \|\zeta^j_{m,2}(t)\|^{2p}\Big] \le C\Big(\int_0^v \mathbb{E}\Big[\sup_{t\in[0,u]} \|\xi_m(t) - Y_m(t)\|^{2p}\Big]du + \delta^{2p}_m\Big),$$

$$\mathbb{E}\Big[\sup_{t\in[0,v]} \|\zeta^j_{m,4}(t)\|^{2p}\Big] \le C\Big(\int_0^v \mathbb{E}\Big[\sup_{t\in[0,u]} \|\xi_m(t) - Y_m(t)\|^{2p}\Big]du + \delta^{2p}_m\Big),$$

$$\mathbb{E}\Big[\sup_{t\in[0,v]} \|\zeta_{m,5}(t)\|^{2p}\Big] \le C\Big(\int_0^v \mathbb{E}\Big[\sup_{t\in[0,u]} \|\xi_m(t) - Y_m(t)\|^{2p}\Big]du + \delta^{p-1}_m\Big).$$

Proof See [20, Lemma 2.6] for the first inequality. For others, see [22, Proposition 4.4, 4.10 and 4.11]. □

In the following, we will prepare lemmas in preparation to estimate the term $\zeta^{j,l}_{m,3}(t)$ for $l \in \{1, \dots, r\}$.

Since we are assuming $(S_t)_{t\ge 0}$ is a C_0-semigroup, the mapping $t \mapsto S_t x$ from $[0, T]$ to H is continuous for all $x \in H$. Furthermore, the following lemma claims that this mapping is even Lipshitz continuous for all $x \in \mathcal{D}$.

Lemma 4.3 *There is a constant $C > 0$ such that for all $t_1, t_2 \in [0, T]$ and all $x \in \mathcal{D}(A)$ we have*

$$\|S_{t_2}x - S_{t_1}x\| \le C\|x\|_{\mathcal{D}(A)}|t_2 - t_1|. \tag{4.5}$$

Proof See [22, Lemma 2.12]. □

The next lemma is an extension of Burkholder's inequality to Hilbert space-valued martingales.

Lemma 4.4 *Let $q > 1$ and $(\mathcal{A}_n)_{n=0,1,\dots}$ be a nondecreasing family of sub σ-fields of $\mathcal{F}$. For any H-valued $(\mathcal{A}_n)_{n=0,1,\dots}$-martingale $(M_n)_{n=0,1,\dots}$ with $M_0 = 0$, there exists a constant $C_q > 0$ depending only on q such that*

$$E\Big[\max_{1\le k\le n} \|M_k\|^{2q}\Big] \le C_q E\Big[\Big(\sum_{k=1}^n \|M_k - M_{k-1}\|^2\Big)^q\Big]$$

for all $n \ge 1$.

Proof See [20, Lemma 2.10]. □

Corollary 4.5 *Let $q > 0$ be a positive real number. Then there is a constant $C > 0$ such that for each $m \in \mathbb{N}$ we have*

$$\mathbb{E}\Big[|B^j((k+1)\delta_m) - B^j(k\delta_m)|^{2q}\Big] \leq C\delta_m^q.$$

Proof See [22, Corollary 4.6] □

Lemma 4.6 *Let $q > 0$ be a positive real number. Then there is a constant $C > 0$ such that for each $m \in \mathbb{N}$ we have*

$$\mathbb{E}\Bigg[\sup_{t\in[0,T]} |\dot{B}_m^j(t)|^{2q}\Bigg] \leq C\delta_m^{-(q+1)}.$$

Proof This can be proven using Corollary 4.5, but for details, see [22, Lemma 4.7]. □

Corollary 4.7 *Let $q > 0$ be a positive real number. Then there is a constant $C > 0$ such that for each $m \in \mathbb{N}$ we have*

$$\mathbb{E}\Bigg[\sup_{t\in[0,T]} \|\xi_m(t)\|_{\mathcal{D}(A)}^{2q}\Bigg] \leq C\delta_m^{-(q+1)}.$$

Proof From Proposition 2.5, $(\xi_m(t))_{t\in[0,T]}$ for each $m \in \mathbb{N}$ is the mild solution on the state space $(\mathcal{D}(A), \|\cdot\|_{\mathcal{D}(A)})$. So, we have

$$\xi_m(t) = S_t x_0 + \big(\mathcal{D}(A)-\big)\int_0^t S_{t-s}\Bigg(b(\xi_m(s)) + \sum_{j=1}^r \sigma_j(\xi_m(s))\dot{B}_m^j(s)\Bigg)ds, \quad t \in [0,T].$$

Here the notation $\big(\mathcal{D}(A)-\big)\int_0^t$ indicates that we consider the respective integral on the state space $(\mathcal{D}(A), \|\cdot\|_{\mathcal{D}(A)})$.

Therefore, taking into account Assumption 2.3, the stated estimate is an immediate consequence of Lemma 4.6. □

Now we estimate $\xi_{m,3}^{j,l}(t)$ for $l \in \{1,\dots,r\}$ and $m \in \mathbb{N}$. For each $\alpha \in (\frac{1}{2p}, \frac{1}{2})$, there is a constant $C > 0$ such that for each $l \in \{1,\dots,r\}$ and each $m \in \mathbb{N}$ we have

$$\mathbb{E}\Bigg[\sup_{t\in[0,T]} \|\zeta_{m,3}^{j,l}(t)\|^{2p}\Bigg] \leq C\delta_m^{p(1-2\alpha)}.$$

This result comes from [20, Lemma 2.11], where Assumption 2.3 is not required. In the following proposition, we get a result whose exponent is any $\kappa \in (0, p)$ instead of $p(1-2\alpha)$ under Assumptions 2.3. Here note that $p(1-2\alpha) < p\left(1-2\frac{1}{2p}\right) = p-1$.

Proposition 4.8 *For each $\kappa \in (0, p)$ there is a constant $C > 0$ such that for each $l \in \{1, \dots, r\}$ and each $m \in \mathbb{N}$ we have*

$$\mathbb{E}\left[\sup_{t\in[0,T]} \|\zeta_{m,3}^{j,l}(t)\|^{2p}\right] \leq C\delta_m^{\kappa}.$$

Proof Let

$$I_m(t) := \delta_m^{-2} \int_0^{[t]_m^-} S_{t-s} D\sigma_j(\bar{\xi}_m(s)) \int_{[s]_m^-}^{s} S_{s-u}\sigma_l(\bar{\xi}_m(u))du K_m(s)ds,$$

where

$$K_m(s) := \begin{cases} (B^j([s]_m^+) - B^j([s]_m^-))^2 - \delta_m, & \text{if } j = l, \\ (B^j([s]_m^+) - B^j([s]_m^-))(B^l([s]_m^+) - B^l([s]_m^-)), & \text{if } j \neq l. \end{cases}$$

Then we have $\zeta_{m,3}^{j,l}(t) = I_m(t)$, and hence, we need to show that

$$\mathbb{E}\left[\sup_{t\in[0,T]} \|I_m(t)\|^{2p}\right] \leq C\delta_m^{\kappa}.$$

Note that

$$I_m(t) = \delta_m^{-2}(I_m^1(t) + I_m^2(t) + I_m^3(t)),$$

where the quantity $I_m^1(t)$ is defined as

$$I_m^1(t) := \int_0^{[t]_m^-} S_{t-s} D\sigma_j(\bar{\xi}_m(s)) \int_{[s]_m^-}^{s} S_{s-u}(\sigma_l(\bar{\xi}_m(u)) - \sigma_l(\xi_m([u]_m^-)))du K_m(s)ds,$$

the quantity $I_m^2(t)$ is defined as

$$I_m^2(t) := \int_0^{[t]_m^-} S_{t-s}(D\sigma_j(\bar{\xi}_m(s)) - D\sigma_j(\xi_m([s]_m^-))) \int_{[s]_m^-}^{s} S_{s-u}\sigma_l(\xi_m([u]_m^-))du K_m(s)ds,$$

and the quantity $I_m^3(t)$ is defined as

$$I_m^3(t) := \int_0^{[t]_m^-} S_{t-s} D\sigma_j(\xi_m([s]_m^-)) \int_{[s]_m^-}^{s} S_{s-u}\sigma_l(\xi_m([u]_m^-)) du K_m(s) ds.$$

Next, we define

$$\tilde{K}_m(k) := \begin{cases} (B^j((k+1)\delta_m) - B^j(k\delta_m))^2 - \delta_m, & \text{if } j = l, \\ (B^j((k+1)\delta_m) - B^j(k\delta_m))(B^l((k+1)\delta_m) - B^l(k\delta_m)), & \text{if } j \neq l, \end{cases}$$

and we choose constants $\pi, \theta \in (1, \infty)$ such that

$$p - \frac{1}{\pi} > \kappa \quad \text{and} \quad \frac{1}{\pi} + \frac{1}{\theta} = 1.$$

Noting that B^j and B^l are independent for $j \neq l$, by Corollary 4.5 (applied with $q = 2p\theta$ in case $j = l$, and applied twice with $q = p\theta$ in case $j \neq l$) we have

$$\mathbb{E}\Big[|\tilde{K}_m(k)|^{2p\theta}\Big]^{1/\theta} \leq C\delta_m^{2p} \quad \text{for all } k = 0, \ldots, m-1.$$

Now, we set

$$U_m(t) := \sup_{0 \leq u \leq \delta_m} \|(S_u - \mathrm{Id})\xi_m(t)\|.$$

By Lemma 4.3, for each $t \in [0, T]$ we have

$$U_m(t) \leq C \sup_{0 \leq u \leq \delta_m} u \|\xi_m(t)\|_{\mathcal{D}(A)} = C\delta_m \|\xi_m(t)\|_{\mathcal{D}(A)}.$$

Therefore, by Corollary 4.7 (applied with $q = p\pi$) we obtain

$$\begin{aligned} \mathbb{E}\Big[\max_{0 \leq k \leq m} U_m(k\delta_m)^{2p\pi}\Big] &\leq \mathbb{E}\Big[\sup_{t \in [0,T]} U_m(t)^{2p\pi}\Big] \leq C\delta_m^{2p\pi}\,\mathbb{E}\Big[\sup_{t \in [0,T]} \|\xi_m(t)\|_{\mathcal{D}(A)}^{2p\pi}\Big] \\ &\leq C\delta_m^{2p\pi}\delta_m^{-(p\pi+1)} = C\delta_m^{p\pi-1}. \end{aligned}$$

Now we have

$$\|I_m^1(t)\| \leq C \int_0^T \int_{[s]_m^-}^{s} \|(S_{u-[s]_m^-} - \mathrm{Id})\xi_m([s]_m^-)\| du |K_m(s)| ds$$

$$\leq C \int_0^T (s - [s]_m^-) U_m([s]_m^-) |K_m(s)| ds$$

$$= C \frac{\delta_m^2}{2} \sum_{k=0}^{m-1} U_m(k\delta_m) |\tilde{K}_m(k)|.$$

Hence, we obtain

$$\|I_m^1(t)\|^{2p} \leq C \left(\frac{\delta_m^2}{2} \right)^{2p} m^{2p-1} \times \sum_{k=0}^{m-1} U_m(k\delta_m)^{2p} |\tilde{K}_m(k)|^{2p}$$

$$\leq C \left(\frac{\delta_m^2}{2} \right)^{2p} m^{2p} \max_{0 \leq k \leq m} U_m(k\delta_m)^{2p} |\tilde{K}_m(k)|^{2p}.$$

By Hölder's inequality we conclude that

$$\mathbb{E}\left[\sup_{t \in [0,T]} \|I_m^1(t)\|^{2p} \right]$$

$$\leq C \left(\frac{\delta_m^2}{2} \right)^{2p} m^{2p} \mathbb{E}\left[\max_{0 \leq k \leq m} U_m(k\delta_m)^{2p\pi} \right]^{1/\pi} \mathbb{E}\left[|\tilde{K}_m(k)|^{2p\theta} \right]^{1/\theta}$$

$$\leq C \delta_m^{4p} \delta_m^{-2p} \delta_m^{p - \frac{1}{\pi}} \delta_m^{2p} \leq C \delta_m^{4p} \delta_m^{\kappa}.$$

Now we have

$$\|I_m^2(t)\| \leq C \int_0^T \|(S_{s-[s]_m^-} - \mathrm{Id}) \xi_m([s]_m^-)\| (s - [s]_m^-) |K_m(s)| ds$$

$$\leq C \int_0^T U_m([s]_m^-)(s - [s]_m^-) |K_m(s)| ds.$$

So, by the same argument as above, we also have

$$\mathbb{E}\left[\sup_{t \in [0,T]} \|I_m^2(t)\|^{2p} \right] \leq C \delta_m^{4p} \delta_m^{\kappa}.$$

Next, we have

$$I_m^3(t) = \sum_{k=0}^{\delta_m^{-1}[t]_m^- - 1} \int_{k\delta_m}^{(k+1)\delta_m} S_{t-s} D\sigma_j(\xi_m(k\delta_m)) \int_{k\delta_m}^{s} S_{s-u} \sigma_l(\xi_m(k\delta_m)) du K_m(s) ds.$$

Let us consider the process $M^m = (M_n^m)_{n=0,1,\ldots,m}$ defined by

$$M_n^m := \sum_{k=0}^{n-1} \int_{k\delta_m}^{(k+1)\delta_m} S_{t-s} D\sigma_j(\xi_m(k\delta_m)) \int_{k\delta_m}^{s} S_{s-u}\sigma_l(\xi_m(k\delta_m)) du K_m(s) ds$$

for $n = 1, \ldots, m$ with $M_0^m = 0$. Then M^m is a $(\mathcal{F}_{n\delta_m})$-martingale, because $\mathbb{E}[K_m(s)] = 0$ for all $s \in [0, T]$, and $\xi_m(k\delta_m)$ is $\mathcal{F}_{k\delta_m}$-measurable and $\tilde{K}_m(k)$ is independent of $\mathcal{F}_{k\delta_m}$ for all $k = 0, \ldots, m-1$. Therefore, from Lemma 4.4, we have

$$\begin{aligned}
\mathbb{E}\left[\sup_{t\in[0,T]} \|I_m^3(t)\|^{2p}\right] &= \mathbb{E}\left[\max_{1\le n\le m} \|M_n^m(t)\|^{2p}\right] \\
&\le C_p \mathbb{E}\left[\left(\sum_{k=0}^{m-1} \|M_{k+1}^m - M_k^m\|^2\right)^p\right] \\
&= C_p \mathbb{E}\Bigg[\Bigg(\sum_{k=0}^{m-1} \Bigg\| \int_{k\delta_m}^{(k+1)\delta_m} S_{t-s} D\sigma_j(\xi_m(k\delta_m)) \\
&\qquad \int_{k\delta_m}^{s} S_{s-u}\sigma_l(\xi_m(k\delta_m)) du K_m(s) ds \Bigg\|^2\Bigg)^p\Bigg],
\end{aligned}$$

where C_p is depending only on p, and hence

$$\begin{aligned}
&\mathbb{E}\left[\sup_{t\in[0,T]} \|I_m^3(t)\|^{2p}\right] \le C\mathbb{E}\left[\left(\sum_{k=0}^{m-1} \left(\int_{k\delta_m}^{(k+1)\delta_m} (s - k\delta_m)|K_m(s)| ds\right)^2\right)^p\right] \\
&= C\left(\frac{\delta_m^2}{2}\right)^{2p} \mathbb{E}\left[\left(\sum_{k=0}^{m-1} |\tilde{K}_m(k)|^2\right)^p\right] \le C\left(\frac{\delta_m^2}{2}\right)^{2p} m^{p-1} \sum_{k=0}^{m-1} \mathbb{E}\left[|\tilde{K}_m(k)|^{2p}\right] \\
&\le C\left(\frac{\delta_m^2}{2}\right)^{2p} m^p \delta_m^{2p} \le C\delta_m^{5p}.
\end{aligned}$$

Combining above results, we have

$$\mathbb{E}\left[\sup_{t\in[0,T]} \|I_m(t)\|^{2p}\right] \le C\delta_m^{-4p}(\delta_m^{4p}\delta_m^{\kappa} + \delta_m^{5p}) = C(\delta_m^{\kappa} + \delta_m^{p}) \le C\delta_m^{\kappa},$$

completing the proof. □

Theorem 2.8 is an immediate consequence of the Eq. (4.1), Propositions 3.3, 4.1, 4.2, and 4.8 and Gronwall's inequality.

5 An Application to the HJM Model

5.1 Framework

Let us consider a SPDE for the HJM model

$$\begin{cases} dr(t) = \dfrac{\partial}{\partial x} r(t)dt + \alpha(r(t), v(t))dt + \displaystyle\sum_{j=1}^{n} \gamma_j(r(t), v(t))dB^j(t), \\ dv(t) = \mu(v(t))dt + \displaystyle\sum_{j=1}^{n} \lambda_j(v(t))dB^j(t), \\ r(0) = r_0, \quad v(0) = v_0, \end{cases} \tag{5.1}$$

which describes forward curves, with a more general volatilities than the volatilities in the SPDE (1.3), where $(B^j(t))_{t\geq 0}$, $j = 1, \ldots, n$ are independent real-valued standard Wiener processes for an integer n with respect to a risk-neutral measure,

$$\alpha(h, v) = \sum_{j=1}^{n} \gamma_j(h, v) \int_0^{\bullet} \gamma_j(h, v)(\eta)d\eta, \tag{5.2}$$

and γ_j is a mapping that takes values on volatility curves and satisfies some appropriate conditions stated below for each $j = 1, \ldots, n$.

In the special case of $v(t) \equiv t$ with $v_0 = 0$, $\mu(v) \equiv 1$ and $\lambda_j(v) \equiv 0$ for all $j = 1, \ldots, n$, the SPDE becomes as follows.

$$\begin{cases} dr(t) = \dfrac{\partial}{\partial x} r(t)dt + \alpha(r(t), t)dt + \displaystyle\sum_{j=1}^{n} \gamma_j(r(t), t)dB^j(t), \\ r(0) = r_0, \end{cases} \tag{5.3}$$

which is called 'local volatility model'. In this case, the volatility curve depends on both the forward curve and the observation time. When $v(t)$ is not deterministic, the model is called 'stochastic volatility model'. These models are used to fit the prices of many option trades, specifically to capture the phenomenon known as the volatility smile.

Hereafter we will formalize the SPDE (5.1) mathematically and apply our Theorem 2.6.

At first, we provide an example concerning the setup of the space for forward curves and the space for volatility curves. Let $\tilde{H}_\beta$ with an arbitrary constant $\beta > 0$ be the space of all absolutely continuous functions $h: \mathbb{R}_+ \to \mathbb{R}$ such that the norm satisfies

$$\|h\|_\beta = \left(|h(0)|^2 + \int_{\mathbb{R}_+} |h'(x)|^2 e^{\beta x} dx \right)^{\frac{1}{2}} < \infty.$$

It holds that the limit $h(\infty) := \lim_{x \to \infty} h(x)$ exists for each $h \in \tilde{H}_\beta$ and the space

$$\tilde{H}_\beta^0 = \{h \in \tilde{H}_\beta;\ h(\infty) = 0\}.$$

is a closed subspace of $\tilde{H}_\beta$ (see [10, Sec. 5.1] for details).

Now let us fix $\beta > 0$ and $\beta' > \beta$. In what follows, we will utilize $\tilde{H}_\beta$ as the space for forward curves and $\tilde{H}_{\beta'}^0$ as the space for volatility curves. It is consistent with intuition that the slopes of the forward curves converge to 0 and the levels of the volatility curves converge to 0 as maturity time x goes to ∞.

The semigroup $(\tilde{S}_t)_{t \ge 0}$ of right shifts in $\tilde{H}_\beta$, defined by $\tilde{S}_t h(x) = h(x+t)$, constitutes a C_0-semigroup with the infinitesimal generator $\tilde{A}$ given by $\tilde{A}h(x) = h'(x)$ on the domain $\mathcal{D}_\beta(\tilde{A}) = \{h \in \tilde{H}_\beta;\ h' \in \tilde{H}_\beta\}$.

Let us assume the following.

Assumption 5.1 We suppose that the following conditions are fulfilled:

- We have $\gamma_j \in C_b^2(\tilde{H}_\beta \times \mathbb{R}; \tilde{H}_{\beta'}^0)$ for $j = 1, \ldots, n$, where $\tilde{H}_\beta \times \mathbb{R}$ is equipped with the norm $\|(h, v)\|_\beta = (\|h\|_\beta^2 + |v|^2)^{\frac{1}{2}}$.
- We have $\gamma_j(\mathcal{D}_\beta(\tilde{A}) \times \mathbb{R}) \subset \mathcal{D}_{\beta'}(\tilde{A})$ and the restriction satisfies $\gamma_j|_{\mathcal{D}_\beta(\tilde{A}) \times \mathbb{R}} \in C_b^2(\mathcal{D}_\beta(\tilde{A}) \times \mathbb{R}; \mathcal{D}_{\beta'}(\tilde{A}))$ for each $j = 1, \ldots, n$ with respect to the corresponding graph norms.
- The function $\mu \colon \mathbb{R} \to \mathbb{R}$ is Lipschitz continuous and bounded.
- We have $\lambda_j \in C_b^2(\mathbb{R})$ for each $j = 1, \ldots, n$.

Let us consider the following SPDE on the separable Hilbert space $H = \tilde{H}_\beta \times \mathbb{R}$ as

$$\begin{cases} dX(t) = (AX(t) + b(X(t)))dt + \sum_{j=1}^{n} \sigma_j(X(t))dB^j(t), \\ X(0) = x_0, \end{cases} \tag{5.4}$$

where

$$x_0 := \begin{pmatrix} r_0 \\ v_0 \end{pmatrix} \in \mathcal{D}_\beta(A) := \mathcal{D}_\beta(\tilde{A}) \times \mathbb{R},$$

$$X(t) := \begin{pmatrix} r(t) \\ v(t) \end{pmatrix} \in H,$$

$$S_t x := \begin{pmatrix} \tilde{S}_t h \\ v \end{pmatrix} \text{ for } x = \begin{pmatrix} h \\ v \end{pmatrix} \in H,$$

$$Ax := \begin{pmatrix} \tilde{A}h \\ 0 \end{pmatrix} \text{ for } x = \begin{pmatrix} h \\ v \end{pmatrix} \in \mathcal{D}(A),$$

$$b := \begin{pmatrix} b^1 \\ b^2 \end{pmatrix} : H \to H, \quad b^1(h, v) = \alpha(h, v), \quad b^2(h, v) = \mu(v),$$

$$\sigma_j := \begin{pmatrix} \sigma_j^1 \\ \sigma_j^2 \end{pmatrix} : H \to H, \quad \sigma_j^1(h, v) = \gamma_j(h, v), \quad \sigma_j^2(h, v) = \lambda_j(v).$$

The Wong-Zakai approximation $(\varrho_m(t), \zeta_m(t))_{t\in[0,T]}$ with $m \in \mathbb{N}$ to the SPDE (5.4) are the solutions to the following integral equations

$$\begin{aligned}\varrho_m(t, x) = \varrho_m(k\delta_m, x + t - k\delta_m) + \int_{k\delta_m}^t \Big(\alpha(\varrho_m(s), \zeta_m(s), x + t - s) \\ - \beta(\varrho_m(s), \zeta_m(s), x + t - s) \\ + \sum_{j=1}^n \frac{B^j((k+1)\delta_m) - B^j(k\delta_m)}{\delta_m} \gamma_j(\varrho_m(s), \zeta_m(s), x + t - s)\Big) ds,\end{aligned}$$

$$\begin{aligned}\zeta_m(t) = \zeta_m(k\delta_m) + \int_{k\delta_m}^t \Big(\mu(\zeta_m(s)) - \frac{1}{2}\sum_{j=1}^n \lambda_j(\zeta_m(s))\lambda_j'(\zeta_m(s)) \\ + \sum_{j=1}^n \frac{B^j((k+1)\delta_m) - B^j(k\delta_m)}{\delta_m} \lambda_j(\zeta_m(s))\Big) ds\end{aligned}$$

for $t \in [k\delta_m, (k+1)\delta_m]$ and $k = 0, \ldots, m-1$, where

$$\beta(r, v) := \frac{1}{2}\sum_{j=1}^n \big(D_r\gamma_j(r, v)\gamma_j(r, v) + D_v\gamma_j(r, v)\lambda_j(v)\big).$$

Here we utilize the fact that for each $x \in [0, \infty)$, the point evaluation operator $h \mapsto h(x) : \tilde{H}_\beta \to \mathbb{R}$ is a continuous linear functional. Therefore, we can interchange the order of the point evaluation operator and the integration.

With the above settings, we can prove that all assumptions for Theorem 2.6, including Assumption 2.2 and Assumption 2.3, are satisfied (see [22, Section 5] for details).

Theorem 5.2 *Suppose that Assumption 5.1 is fulfilled, and let $T > 0$, $p > 1$ be arbitrary. Then there is a constant $C > 0$ such that for each $m \in \mathbb{N}$ we have*

$$\mathbb{E}\left[\sup_{t\in[0,T]} \left(\|\varrho_m(t) - r(t)\|_\beta^2 + |\zeta_m(t) - v(t)|^2\right)^p\right] \le \frac{C}{m^{p-1}}. \tag{5.5}$$

References

1. Acquistapace, P., Terreni, B.: An approach to Itô linear equations in Hilbert spaces by approximation of white noise with coloured noise. Stoch. Anal. Appl. **2**(2), 131–186 (1984)
2. Aida, S.: Support theorem for diffusion processes on Hilbert spaces. Publ. Res. Inst. Math. Sci. **26**(6), 947–965 (1990)
3. Björk, T., Christensen, B.J.: Interest rate dynamics and consistent forward rate curves. Math. Financ. **9**(4), 323–348 (1999)
4. Brace, A., Musiela, M.: A multifactor Gauss Markov implementation of Heath, Jarrow, and Morton. Math. Financ. **4**(3), 259–283 (1994)
5. Brzeźniak, Z., Carroll, A.: Approximations of the Wong-Zakai type for stochastic differential equations in M-type 2 Banach spaces with applications to loop spaces. In *Séminaire de Probabilités XXXVII, Number 1832 in Lecture Notes in Mathematics*, pp. 251–289. Springer, Berlin (2003)
6. Brzeźniak, Z., Flandoli, F.: Almost sure approximation of Wong-Zakai type for stochastic partial differential equations. Stoch. Process. Appl. **55**(2), 329–358 (1995)
7. Cao, W., Zhang, Z., Karniadakis, G.: Numerical methods for stochastic delay differential equations via the Wong-Zakai approximation. SIAM J. Sci. Comput. **37**(1), A295–A318 (2015)
8. Carmona,R̃., Tehranchi, M.: Interest Rate Models: An Infinite Dimensional Stochastic Analysis Perspective. Springer Finance. Springer, Berlin (2006)
9. Da Prato, G., Zabczyk, J.: Stochastic Equations in Infinite Dimensions, 2nd edn. Cambridge University Press, New York (2014)
10. Filipović, D.: Consistency Problems for Heath-Jarrow-Morton Interest Rate Models. Springer, Berlin (2001)
11. Filipović, D., Teichmann, J.: On the geometry of the term structure of interest rates. In: Proceedings of the Royal Society of London. Series A. Mathematical, Physical and Engineering Sciences **460**, 129–167 (2004)
12. Ganguly, A.: Wong-Zakai type convergence in infinite dimensions. Electron. J. Probab. **18**(31), 1–34 (2013)
13. Gyöngy, I., Shmatkov, A.: Rate of convergence of Wong-Zakai approximations for stochastic partial differential equations. Appl. Math. Optim. **54**(3), 315–341 (2006)
14. Gyöngy, I., Stinga, P.R.: Rate of convergence of Wong-Zakai approximations for stochastic partial differential equations. In: Seminar on Stochastic Analysis, Random Fields and Applications VII, vol. 67, pp. 95–130. Springer, Basel (2013)
15. Hairer, M., Pardoux, E.: A Wong-Zakai theorem for stochastic PDEs. J. Math. Soc. Jpn. **67**(4), 1551–1604 (2015)
16. Hausenblas, E.: Wong-Zakai type approximation of SPDEs of Lévy noise. Acta Appl. Math. **98**(2), 99–134 (2007)
17. Heath, D., Jarrow, R., Morton, A.: Bond pricing and the term structure of interest rates: a new methodology for contingent claims valuation. Econometrica **60**(1), 77–105 (1992)
18. Kamrani, M., Jamshidi, N.: Implicit Milstein method for stochastic differential equations via the Wong-Zakai approximation. Numer. Algorithms **79**(2), 357–374 (2018)
19. Londoño, J.A., Villegas, A.M.: Numerical performance of some Wong-Zakai type approximations for stochastic differential equations. Int. J. Pure Appl. Math. **107**(2), 301–315 (2016)
20. Nakayama, T.: Support theorem for mild solutions of SDE's in Hilbert spaces. J. Math. Sci. Univ. Tokyo **11**(3), 245–311 (2004)
21. Nakayama, T.: Viability Theorem for SPDE's including HJM framework. J. Math. Sci. Univ. Tokyo **11**(3), 313–324 (2004)
22. Nakayama, T., Tappe, S.: Wong-Zakai approximations with convergence rate for stochastic partial differential equations. Stoch. Anal. Appl. **36**(5), 832–857 (2018)
23. Pazy, A.: Semigroups of linear operators and applications to partial differential equations. Springer, New York (1983)

24. Tessitore, G., Zabczyk, J.: Wong-Zakai approximations of stochastic evolution equations. J. Evol. Equ. **6**(4), 621–655 (2006)
25. Twardowska, K.: An extension of the Wong-Zakai theorem for stochastic evolution equations in Hilbert spaces. Stoch. Anal. Appl. **10**(4), 471–500 (1992)
26. Twardowska, K.: Approximation theorems of Wong-Zakai type for stochastic differential equations in infinite dimensions. PhD Thesis, Warsaw Technical University, Institute of Mathematics (1993)
27. Twardowska, K.: An approximation theorem of Wong-Zakai type for nonlinear stochastic partial differential equations. Stoch. Anal. Appl. **13**(5), 601–626 (1995)
28. Twardowska, K.: An approximation theorem of Wong-Zakai type for stochastic Navier-Stokes equations. Rend. Sem. Mat. Univ. Padova **96**, 15–36 (1996)
29. Twardowska, K.: Wong-Zakai approximations for stochastic differential equations. Acta Appl. Math. **43**(3), 317–359 (1996)
30. Wong, E., Zakai, M.: On the relation between ordinary and stochastic differential equations. Int. J. Eng. Sci. **3**(2), 213–229 (1965)

On a Prolongation of the Nonlinear Stochastic Asymptotic Expansion of the Solution of a Semilinear PDE

Sae Odaira and Kaori Okuma

Abstract The present paper studies an error estimate for an approximation of the higher order derivatives of the solution of a semi-linear partial differential equation. A numerical scheme that requires the estimate is introduced, together with reviews of literature.

Keywords Backward stochastic differential equation · Semi-linear partial differential equation · Non-linear Clark–Ocone formula · Deep learning · Asymptotic expansion

1 Introduction

Let $T > 0$ be fixed and W be a d-dimensional Wiener process on a probability space $(\Omega, \mathcal{F}, \mathbb{P})$ equipped with a filtration $\mathbb{F} = (\mathcal{F}_t)_{0 \le t \le T}$ satisfying the usual conditions. For the solution of forward-backward stochastic differential equation(FBSDE); for $0 \le t \le T$

$$\begin{cases} dX_t = \mu(X_t)dt + \sigma(X_t)dW_t \\ Y_t = \Phi(X_T) + \int_t^T f(s, X_s, Y_s, Z_s)ds - \int_t^T Z_s^\top dW_s, \end{cases} \tag{1}$$

and the solution $u(t, x)$ of the partial differential equation(PDE)

$$\begin{cases} \partial_t u + \frac{1}{2}\mathrm{tr}\sigma\sigma^\top\nabla^{\otimes 2}u + \mu^\top\nabla u + f(\cdot, \cdot, u, \sigma^\top\nabla u) = 0, \text{ on } [0, T) \times \mathbf{R}^d, \\ u(T, \cdot) = \Phi, \text{ on } \mathbf{R}^d, \end{cases} \tag{2}$$

S. Odaira · K. Okuma (✉)
Ritsumeikan University, Kusatsu, Japan

J. Akahori et al. (eds.), *Recent Developments in Stochastic Numerics and Computational Finance*, ICIAM2023 Springer Series 6,
https://doi.org/10.1007/978-981-95-0652-1_5

where T denotes transposition of matrices/vectors, and we assume that $\sigma : \mathbf{R}^d \to \mathbf{R}^d \times \mathbf{R}^d$, $\mu : \mathbf{R}^d \to \mathbf{R}^d$, $\Phi : \mathbf{R}^d \to \mathbf{R}$, and $f : [0, T] \times \mathbf{R}^d \times \mathbf{R} \times \mathbf{R}^d \to \mathbf{R}$ have some regularity to ensure that the solution u of (2) is smooth, the following relationship holds (see e.g. [10]):

$$Y_t = u(t, X_t),\ 0 \le t \le T$$

and

$$Z_t = \sigma^{\mathsf{T}}(t, X_t)\nabla u(t, X_t),\ 0 \le t \le T.$$

We are interested in obtaining a numerical value of $u(t, x)$ by discrediting the FBSDE (1). Here we give a very quick review of the literature.

For a given interval $[t, T]$, let $t = t_0 < t_1 < \cdots < t_N = T$ be its discretization. All the schemes listed below start with setting $u^N(t_N, \cdot) = g$ and calculate $u^N(t_k, \cdot)$ using $u^N(t_{k+1}, \cdot)$ for $k = N-1, N-2, \cdots, 0$. Approximation of the forward equation (stochastic differential equation for X) is usually discussed separately. To concentrate on the discretization of Y, we assume that $\mu = 0$ and $\sigma = I$ so that $X = W$.

1. First order explicit scheme [3]:

$$Z^N_{t_N}(x) = \nabla\Phi(x), \tag{3}$$

and for $k = N-1, N-2, \cdots, 0$,

$$\begin{aligned}
u^N(t_k, x) &\equiv Y^N_{t_k}(x) \\
&= E[u^N(t_{k+1}, W_{t_{k+1}}) \\
&\qquad + f(t_{k+1}, W_{t_{k+1}}, Y^N_{t_{k+1}}, Z^N_{t_{k+1}})(t_{k+1} - t_k) | W_{t_k} = x],
\end{aligned}$$

$$\begin{aligned}
&Z^N_{t_k}(x) \\
&= E[(u^N(t_{k+1}, W_{t_{k+1}}) + f(t_{k+1}, W_{t_{k+1}}, Y^N_{t_{k+1}}, Z^N_{t_{k+1}})(t_{k+1} - t_k)) \\
&\qquad \times \frac{W_{t_{k+1}} - W_{t_k}}{t_{k+1} - t_k} | W_{t_k} = x].
\end{aligned} \tag{4}$$

2. First order implicit scheme: for $k = N-1, N-2, \cdots, 0$,

$$\begin{aligned}
u^N(t_k, x) &\equiv Y^N_{t_k}(x) \\
&= E[u^N(t_{k+1}, W_{t_{k+1}}) | W_{t_k} = x] + f(t_k, W_{t_k}, Y^N_{t_k}, Z^N_{t_k})(t_{k+1} - t_k), \\
Z^N_{t_k}(x) &= E[u^N(t_{k+1}, W_{t_{k+1}}) \frac{W_{t_{k+1}} - W_{t_k}}{t_{k+1} - t_k} | W_{t_k} = x].
\end{aligned} \tag{5}$$

3. Second order mixed scheme [7]: for $k = N-1, N-2, \cdots, 0$,

$$
\begin{aligned}
u^N(t_k, x) &\equiv Y_{t_k}^N(x) \\
&= E[Y_{t_{k+1}} + \frac{1}{2} f(t_{k+1}, W_{t_{k+1}}, Y_{t_{k+1}}^N, Z_{t_{k+1}}^N)(t_{k+1} - t_k) | \mathcal{F}_{t_k}] \\
&\quad + \frac{1}{2} f(t_k, W_{t_k}, Y_{t_k}^N, Z_{t_k}^N)(t_{k+1} - t_k).
\end{aligned}
$$

(The calculation of Z^N is involved and we omit it.)

The order of the above schemes largely determined by the approximation scheme of the integral of f. The third scheme relies on the trapezoid rule, and there are also higher order approximation schemes based on Romberg's method [5], or Ruge-Kutta method (see e.g. [4]). The approximation of the sothcastic integral of course needs an order-made approach. See for example [11]. For more details of the literature we refer to the elaborated survey [6].

Those schemes avoid computing $\partial_x u(t_k, X_k)$ (except for (3)), but use *integration* instead in (4) and (5). This is mainly because the numerical calculation of derivatives is generally unstable.

If the derivatives $\partial_x^n u(t, x)$ were available, we may well use a Taylor series expansion to obtain a higher order approximation. In [13], this idea is employed, for the case $d = 1$ for clarity, combined with the nonlinear Feynman-Kac formula. The obtained expansion involves the higher order derivatives, but how they can be numerically obtained in each step is not explained in [13]. In the present paper we will propose a scheme to numerically obtain them; they are given by the conditional expectation of the n-th derivative, $E[\partial_x^n \tilde{u}(t+\Delta, W_{t+\Delta})|\mathcal{F}_t]$, where $\tilde{u}$ is the approximation of u at $t + \Delta$ obtained via the scheme. They can be numerically calculated by solving an optimization problem for neural network functions—possibly by *deep learning*— via the non-linear Clark-Ocone formula proposed in [2]. Details will be given in Sect. 3.

In the present paper, though we still restrict ourselves to the case where $d = 1$, $\mu = 0$ and $\sigma = 1$ for simplicity, we obtain, as the main results, the error estimates $|\partial_x^n u(t, W_t) - E[\partial_x^n \tilde{u}(t+\Delta, W_{t+\Delta})|\mathcal{F}_t]|$ of the second- and the third order.

Let $C_p^\infty([0, T] \times \mathbf{R})$ be the set of C^∞ functions whose derivatives of any order (including the 0-th one) are all with at most polynomial growth in x and bounded in t. For $w \in C_p^\infty([0, T] \times \mathbf{R})$, we denote by $\partial_1 w$ and $\partial_2 w$ the partial derivative with respect to the first variable, and the second one, respectively. Similarly, for $f \in C_p^\infty([0, T] \times \mathbf{R}^3)$, we denote by $\partial_1 f$, $\partial_2 f$, $\partial_3 f$ and $\partial_4 f$ the partial derivative with respect to the first variable, the second one, the third one, and the fourth one respectively. Let $\mathcal{L} := \partial_1 + \frac{1}{2}\partial_2^2$.

Theorem 1 *Let $\Phi \in L^2(\mathbf{R}, \mu)$ where*

$$
\mu(dx) = \frac{1}{(2\pi\Delta t)^{\frac{1}{2}}} e^{-\frac{x^2}{2\Delta t}} dx,
$$

and $f \in C_p^\infty([0,T] \times \mathbf{R}^3)$. *We assume that the solution* u *of the equation*

$$\mathcal{L}u(t,x) = -f(t,x,u(t,x),\partial_2 u(t,x)), \quad \Phi(W_T) = u(T, W_T) \tag{6}$$

exists in $C_p^\infty([0,T] \times \mathbf{R})$ *for* $t \in [T - \Delta t, T]$ *for* $T > 0$ *and* $0 < \Delta t < T$. *Then, for* $k = 2, 3$, *there is a function* $C_k \in C_p^\infty(\mathbf{R})$ *such that*

$$\begin{aligned} &|E[\partial_x^n \Phi(W_T) + G_{n,k}(T, W_T)|\mathcal{F}_{T-\Delta t}] - \partial_2^n u(T - \Delta t, W_{T-\Delta t})| \\ &\leq C_k(W_{T-\Delta t})(\Delta t)^k, \end{aligned}$$

where the differential $\partial_x^n \Phi(x)$ *is in the sense of the distribution. Here*

$$G_{n,2}(T, W_T) = (\Delta t)\partial_2^n f(T, W_T, u(T, W_T), \partial_2 u(T, W_T))$$

and

$$G_{n,3}(T, W_T) = G_{n,2}(T, W_T) - \frac{(\Delta t)^2}{2}\mathcal{L}\partial_2^n f(T, W_T, u(T, W_T), \partial_2 u(T, W_T)).$$

A proof will be given in Sect. 2. The meaning of Theorem 1 in our proposed scheme will then be discussed in Sect. 3, after recalling the setting and the results of [1, 13], and [2]. Section 4 concludes the present paper.

2 The Proof of Theorem 1

Proof Let $H_n(t,x)(n \in \mathbf{Z}_+)$ be the Hermite polynomial of the following form,

$$H_n(t,x) = (-1)^n e^{\frac{x^2}{2t}} \frac{d^n}{dx^n} e^{-\frac{x^2}{2t}}.$$

Then, applying the integration by parts in the distribution sense to $E[\partial_x^n \Phi(W_T)|\mathcal{F}_{T-\Delta t}]$, we obtain

$$E[\partial_x^n \Phi(W_T)|\mathcal{F}_{T-\Delta t}] = E[H_n(\Delta t, \Delta W_T)\Phi(W_T)|\mathcal{F}_{T-\Delta t}]$$

where $\Delta W_T = W_T - W_{T-\Delta t}$. Since

$$\begin{aligned} u(T - \Delta t, W_{T-\Delta t}) = \Phi(W_T) &+ \int_{T-\Delta t}^{T} f(s, W_s, u(s, W_s), \partial_2 u(s, W_s))ds \\ &- \int_{T-\Delta t}^{T} \partial_2 u(s, W_s)dW_s, \end{aligned}$$

we have

$$E[H_n(\Delta t, \Delta W_T)\Phi(W_T)|\mathcal{F}_{T-\Delta t}] = E[H_n(\Delta t, \Delta W_T)\Big(u(T-\Delta t, W_{T-\Delta t}) - \int_{T-\Delta t}^{T} f(s, W_s, u(s, W_s), \partial_2 u(s, W_s))ds + \int_{T-\Delta t}^{T} \partial_2 u(s, W_s)dW_s\Big)|\mathcal{F}_{T-\Delta t}].$$

Moreover, since $u(T-\Delta t, W_{T-\Delta t})$ is $\mathcal{F}_{T-\Delta t}$-measurable and

$$E[H_n(\Delta t, \Delta W_T)|\mathcal{F}_{T-\Delta t}] = 0,$$

we have

$$\begin{aligned} &E[\partial_x^n \Phi(W_T)|\mathcal{F}_{T-\Delta t}] \\ &= E[H_n(\Delta t, \Delta W_T)\Big(- \int_{T-\Delta t}^{T} f(s, W_s, u(s, W_s), \partial_2 u(s, W_s))ds \\ &+ \int_{T-\Delta t}^{T} \partial_2 u(s, W_s)dW_s\Big)|\mathcal{F}_{T-\Delta t}]. \end{aligned}$$

By the n times of integration by parts, we see that

$$\begin{aligned} &E[-\int_{T-\Delta t}^{T} H_n(\Delta t, \Delta W_T) f(s, W_s, u(s, W_s), \partial_2 u(s, W_s))ds|\mathcal{F}_{T-\Delta t}] \\ &= E[-\int_{T-\Delta t}^{T} \frac{(s-(T-\Delta t))^n}{(\Delta t)^n} \partial_2^n f(s, W_s, u(s, W_s), \partial_2 u(s, W_s))ds|\mathcal{F}_{T-\Delta t}]. \end{aligned} \tag{7}$$

To calculate

$$E[H_n(\Delta t, \Delta W_T)\Big(\int_{T-\Delta t}^{T} \partial_2 u(s, W_s)dW_s\Big)|\mathcal{F}_{T-\Delta t}],$$

let us recall some basic facts about the Hermite polynomials $H_n(t, x)$. Let $\tilde{H}_n(t, x) := t^n H_n(t, x)$ and consider its generating function $G(t, x, \beta) := e^{\beta x - \frac{\beta^2 t}{2}}$;

$$G(t, x, \beta) = \sum_{n=0}^{\infty} \frac{\beta^n}{n!} t^n H_n(t, x) = \sum_{n=0}^{\infty} \frac{\beta^n}{n!} \tilde{H}_n(t, x).$$

Since the generating function $G(t, x, \beta)$ is a time-space harmonic, i.e., $(\partial_t + \frac{1}{2}\partial_x^2)G(t, x, \beta) = 0$, we have

$$\sum_{n=0}^{\infty} \frac{\beta^n}{n!}(\partial_t + \frac{1}{2}\partial_x^2)\tilde{H}_n(t, x) = 0.$$

The above equation implies that each $\tilde{H}_n(t,x)$ also is a time-space harmonic;

$$\partial_t \tilde{H}_n(t,x) + \frac{1}{2}\partial_x^2 \tilde{H}_n(t,x) = 0.$$

Noting that

$$\begin{aligned} &E[H_n(\Delta t, \Delta W_T)\int_{T-\Delta t}^{T} \partial_2 u(s, W_s)dW_s|\mathcal{F}_{T-\Delta t}] \\ &= \frac{1}{(\Delta t)^n} E[\tilde{H}_n(\Delta t, \Delta W_T)\int_{T-\Delta t}^{T} \partial_2 u(s, W_s)dW_s|\mathcal{F}_{T-\Delta t}], \end{aligned}$$

we apply Itô's formula to $\tilde{H}_n(T-(T-\Delta t), W_T - W_{T-\Delta t})$ to obtain

$$\begin{aligned} &\tilde{H}_n(T-(T-\Delta t), W_T - W_{T-\Delta t}) \\ &= \tilde{H}_n(0,0) + \int_{T-\Delta t}^{T} \partial_x \tilde{H}_n(s-(T-\Delta t), W_s - W_{T-\Delta t})dW_s \\ &+ \int_{T-\Delta t}^{T} (\partial_t + \frac{1}{2}\partial_x^2)\tilde{H}_n(s-(T-\Delta t), W_s - W_{T-\Delta t})ds. \end{aligned}$$

Since $\tilde{H}_n(0,0) = 0$ and $\tilde{H}_n(t,x)$ is a time-space harmonic,

$$\begin{aligned} &\tilde{H}_n(T-(T-\Delta t), W_T - W_{T-\Delta t}) \\ &= \int_{T-\Delta t}^{T} \partial_x \tilde{H}_n(s-(T-\Delta t), W_s - W_{T-\Delta t})dW_s. \end{aligned}$$

For the Hermite polynomial $H_n(t,x)$, it holds that

$$\partial_x H_n(t,x) = \frac{n}{t} H_{n-1}(t,x),$$

which is,

$$\partial_x \tilde{H}_n(t,x) = t^{n-1} n H_{n-1}(t,x).$$

Therefore,

$$\begin{aligned} &\tilde{H}_n(\Delta t, \Delta W_T) \\ &= \int_{T-\Delta t}^{T} (s-(T-\Delta t))^{n-1} n H_{n-1}(s-(T-\Delta t), W_s - W_{T-\Delta t})dW_s, \end{aligned}$$

and thus

$$
\begin{aligned}
&E[H_n(\Delta t, \Delta W_T)\int_{T-\Delta t}^{T} \partial_2 u(s, W_s)dW_s|\mathcal{F}_{T-\Delta t}] \\
&= \frac{1}{(\Delta t)^n}E[\int_{T-\Delta t}^{T} (s-(T-\Delta t))^{n-1} nH_{n-1}(s-(T-\Delta t), W_s - W_{T-\Delta t})dW_s \\
&\qquad \int_{T-\Delta t}^{T} \partial_2 u(s, W_s)dW_s|\mathcal{F}_{T-\Delta t}].
\end{aligned}
$$

Then by Itô isometry,

$$
\begin{aligned}
&E[H_n(\Delta t, \Delta W_T)\Big(\int_{T-\Delta t}^{T} \partial_2 u(s, W_s)dW_s\Big)|\mathcal{F}_{T-\Delta t}] \\
&= \frac{1}{(\Delta t)^n}E[\int_{T-\Delta t}^{T} (s-(T-\Delta t))^{n-1} nH_{n-1}(s-(T-\Delta t), W_s - W_{T-\Delta t}) \\
&\qquad \partial_2 u(s, W_s)ds|\mathcal{F}_{T-\Delta t}].
\end{aligned}
$$

We can exchange the order of the expectation and the integral and we apply the integration by parts to $E[H_{n-1}(s-(T-\Delta t), W_s - W_{T-\Delta t})\partial_2 u(s, W_s)|\mathcal{F}_{T-\Delta t}]$ to obtain

$$
\begin{aligned}
&\frac{1}{(\Delta t)^n}\int_{T-\Delta t}^{T} (s-(T-\Delta t))^{n-1} \\
&\quad nE[H_{n-1}(s-(T-\Delta t), W_s - W_{T-\Delta t})\partial_2 u(s, W_s)|\mathcal{F}_{T-\Delta t}]ds \\
&\quad = \frac{1}{(\Delta t)^n}\int_{T-\Delta t}^{T} (s-(T-\Delta t))^{n-1} nE[\partial_2^n u(s, W_s)|\mathcal{F}_{T-\Delta t}]ds.
\end{aligned}
$$

In summary, we obtained

$$
\begin{aligned}
&E[H_n(\Delta t, \Delta W_T)\Big(\int_{T-\Delta t}^{T} \partial_2 u(s, W_s)dW_s\Big)|\mathcal{F}_{T-\Delta t}] \\
&= \frac{1}{(\Delta t)^n}\int_{T-\Delta t}^{T} (s-(T-\Delta t))^{n-1} nE[\partial_2^n u(s, W_s)|\mathcal{F}_{T-\Delta t}]ds,
\end{aligned}
\tag{8}
$$

and by (7) and (8), we have

$$
\begin{aligned}
&E[\partial_x^n \Phi(W_T)|\mathcal{F}_{T-\Delta t}] \\
&= -E[\int_{T-\Delta t}^{T} \frac{(s-(T-\Delta t))^n}{(\Delta t)^n}\partial_2^n f(s, W_s, u(s, W_s), \partial_2 u(s, W_s))ds|\mathcal{F}_{T-\Delta t}] \\
&\quad + \frac{1}{(\Delta t)^n}\int_{T-\Delta t}^{T} (s-(T-\Delta t))^{n-1} nE[\partial_2^n u(s, W_s)|\mathcal{F}_{T-\Delta t}]ds.
\end{aligned}
\tag{9}
$$

Let $g(s, W_s) := \partial_2^n f(s, W_s, u(s, W_s), \partial_2 u(s, W_s))$. Applying Itô's formula to $g(s, W_s)$, we see that

$$
\begin{aligned}
&E[g(s, W_s)|\mathcal{F}_{T-\Delta t}] \\
&= E[\Big(g(T-\Delta t, W_{T-\Delta t}) + \int_{T-\Delta t}^{s} \partial_1 g(v_1, W_{v_1}) dv_1 \\
&\quad + \int_{T-\Delta t}^{s} \partial_2 g(v_1, W_{v_1}) dW_{v_1} + \frac{1}{2}\int_{T-\Delta t}^{s} \partial_2^2 g(v_1, W_{v_1}) dv_1\Big)|\mathcal{F}_{T-\Delta t}].
\end{aligned}
$$

Since $g(T-\Delta t, W_{T-\Delta t})$ is $\mathcal{F}_{T-\Delta t}$-measurable and

$$
E[\int_{T-\Delta t}^{s} \partial_2 g(v_1, W_{v_1}) dW_{v_1}|\mathcal{F}_{T-\Delta t}] = 0,
$$

we see that

$$
\begin{aligned}
&E[g(s, W_s)|\mathcal{F}_{T-\Delta t}] \\
&= g(T-\Delta t, W_{T-\Delta t}) + E[\int_{T-\Delta t}^{s} \mathcal{L}g(v_1, W_{v_1}) dv_1|\mathcal{F}_{T-\Delta t}].
\end{aligned}
$$

Appling Itô's formula to $\mathcal{L}g(v_1, W_{v_1})$,

$$
\begin{aligned}
&E[g(s, W_s)|\mathcal{F}_{T-\Delta t}] \\
&= g(T-\Delta t, W_{T-\Delta t}) \\
&\quad + E[\int_{T-\Delta t}^{s} \Big(\mathcal{L}g(T-\Delta t, W_{T-\Delta t}) + \int_{T-\Delta t}^{v_1} \mathcal{L}^2 g(v_2, W_{v_2}) dv_2 \\
&+ \int_{T-\Delta t}^{v_1} \partial_2 \mathcal{L}g(v_2, W_{v_2}) dW_{v_2}\Big) dv_1|\mathcal{F}_{T-\Delta t}] \\
&= g(T-\Delta t, W_{T-\Delta t}) + E[\int_{T-\Delta t}^{s} \mathcal{L}g(T-\Delta t, W_{T-\Delta t}) dv_1|\mathcal{F}_{T-\Delta t}] \\
&+ E[\int_{T-\Delta t}^{s} \int_{T-\Delta t}^{v_1} \mathcal{L}^2 g(v_2, W_{v_2}) dv_2 dv_1|\mathcal{F}_{T-\Delta t}].
\end{aligned}
$$

Thus, we obtain

$$
\begin{aligned}
&E[g(s, W_s)|\mathcal{F}_{T-\Delta t}] \\
&= g(T-\Delta t, W_{T-\Delta t}) + (s-(T-\Delta t))\mathcal{L}g(T-\Delta t, W_{T-\Delta t}) \\
&+ E[\int_{T-\Delta t}^{s} \int_{T-\Delta t}^{v_1} \mathcal{L}^2 g(v_2, W_{v_2}) dv_2 dv_1|\mathcal{F}_{T-\Delta t}].
\end{aligned}
\tag{10}
$$

Similarly, by applying Itô's formula to $\partial_2^n u(s, W_s)$, we have

$$
\begin{aligned}
&E[\partial_2^n u(s, W_s)|\mathcal{F}_{T-\Delta t}] \\
&= \partial_2^n u(T-\Delta t, W_{T-\Delta t}) + E[\int_{T-\Delta t}^{s} \mathcal{L}\partial_2^n u(v_1, W_{v_1}) dv_1 |\mathcal{F}_{T-\Delta t}].
\end{aligned}
$$

Applying Itô's formula to $\mathcal{L}\partial_2^n u(v_1, W_{v_1})$, and we can rewrite $E[\partial_2^n u(s, W_s)|\mathcal{F}_{T-\Delta t}]$ as follows:

$$
\begin{aligned}
&E[\partial_2^n u(s, W_s)|\mathcal{F}_{T-\Delta t}] \\
&= \partial_2^n u(T-\Delta t, W_{T-\Delta t}) + E[\int_{T-\Delta t}^{s} \Big(\mathcal{L}\partial_2^n u(T-\Delta t, W_{T-\Delta t}) \\
&\quad + \int_{T-\Delta t}^{v_1} \mathcal{L}^2 \partial_2^n u(v_2, W_{v_2}) dv_2 + \int_{T-\Delta t}^{v_1} \partial_2 \partial_2^n u(v_2, W_{v_2}) dW_{v_2} \Big) dv_1 |\mathcal{F}_{T-\Delta t}] \\
&= \partial_2^n u(T-\Delta t, W_{T-\Delta t}) + \mathcal{L}\partial_2^n u(T-\Delta t, W_{T-\Delta t})(s-(T-\Delta t)) \\
&\qquad + E[\int_{T-\Delta t}^{s} \int_{T-\Delta t}^{v_1} \mathcal{L}^2 \partial_2^n u(v_2, W_{v_2}) dv_2 dv_1 |\mathcal{F}_{T-\Delta t}].
\end{aligned}
$$

Now, by applying Itô's formula to $\mathcal{L}^2\partial_2^n u(v_2, W_{v_2})$, we have

$$
\begin{aligned}
&\partial_2^n u(T-\Delta t, W_{T-\Delta t}) + \mathcal{L}\partial_2^n u(T-\Delta t, W_{T-\Delta t})(s-(T-\Delta t)) \\
&+ E[\int_{T-\Delta t}^{s} \int_{T-\Delta t}^{v_1} \Big(\mathcal{L}^2 \partial_2^n u(T-\Delta t, W_{T-\Delta t}) + \int_{T-\Delta t}^{v_2} \mathcal{L}^3 \partial_2^n u(v_3, W_{v_3}) dv_3 \\
&\qquad + \int_{T-\Delta t}^{v_2} \partial_2 \mathcal{L}^2 \partial_2^n u(v_3, W_{v_3}) dW_{v_3} \Big) dv_2 dv_1 |\mathcal{F}_{T-\Delta t}].
\end{aligned}
$$

Since $\mathcal{L}^2\partial_2^n u(T-\Delta t, W_{T-\Delta t})$ is $\mathcal{F}_{T-\Delta t}$-measurable and

$$
E[\int_{T-\Delta t}^{s} \int_{T-\Delta t}^{v_1} \Big(\int_{T-\Delta t}^{v_2} \partial_2 \mathcal{L}^2 \partial_2^n u(v_3, W_{v_3}) dW_{v_3} \Big) dv_2 dv_1 |\mathcal{F}_{T-\Delta t}] = 0,
$$

we have

$$
\begin{aligned}
&E[\partial_2^n u(s, W_s)|\mathcal{F}_{T-\Delta t}] \\
&= \partial_2^n u(T-\Delta t, W_{T-\Delta t}) + \mathcal{L}\partial_2^n u(T-\Delta t, W_{T-\Delta t})(s-(T-\Delta t)) \\
&+ \frac{1}{2}(s-(T-\Delta t))^2 \mathcal{L}^2 \partial_2^n u(T-\Delta t, W_{T-\Delta t}) \qquad (11) \\
&+ E[\int_{T-\Delta t}^{s} \int_{T-\Delta t}^{v_1} \int_{T-\Delta t}^{v_2} \mathcal{L}^3 \partial_2^n u(v_3, W_{v_3}) dv_3 dv_2 dv_1 |\mathcal{F}_{T-\Delta t}].
\end{aligned}
$$

We will consider $I_n := E[\partial_x^n \Phi(W_T)|\mathcal{F}_{T-\Delta t}] - \partial_2^n u(T-\Delta t, W_{T-\Delta t})$. By (9),

$$I_n = -\int_{T-\Delta t}^{T} \frac{(s-(T-\Delta t))^n}{(\Delta t)^n} E[\partial_2^n f(s, W_s, u(s, W_s), \partial_2 u(s, W_s))|\mathcal{F}_{T-\Delta t}]ds$$
$$+ \frac{1}{(\Delta t)^n}\int_{T-\Delta t}^{T} (s-(T-\Delta t))^{n-1} n E[\partial_2^n u(s, W_s)|\mathcal{F}_{T-\Delta t}]ds$$
$$- \partial_2^n u(T-\Delta t, W_{T-\Delta t}). \tag{12}$$

Substituting (10) and (11) into the above Eq. (12), we obtain

$$\begin{aligned}
I_n = &-\int_{T-\Delta t}^{T} \frac{(s-(T-\Delta t))^n}{(\Delta t)^n}\Big(g(T-\Delta t, W_{T-\Delta t})\\
&+ (s-(T-\Delta t))\mathcal{L}g(T-\Delta t, W_{T-\Delta t})\\
&+ E[\int_{T-\Delta t}^{s}\int_{T-\Delta t}^{v_1} \mathcal{L}^2 g(v_2, W_{v_2})dv_2 dv_1|\mathcal{F}_{T-\Delta t}]\Big)ds\\
&\quad + \frac{1}{(\Delta t)^n}\int_{T-\Delta t}^{T}(s-(T-\Delta t))^{n-1} n\Big(\partial_2^n u(T-\Delta t, W_{T-\Delta t})\\
&+ (s-(T-\Delta t))\mathcal{L}\partial_2^n u(T-\Delta t, W_{T-\Delta t})\\
&+ \frac{1}{2}(s-(T-\Delta t))^2\mathcal{L}^2\partial_2^n u(T-\Delta t, W_{T-\Delta t})\\
&+ E[\int_{T-\Delta t}^{s}\int_{T-\Delta t}^{v_1}\int_{T-\Delta t}^{v_2} \mathcal{L}^3\partial_2^n u(v_3, W_{v_3})dv_3|\mathcal{F}_{T-\Delta t}]\Big)ds\\
&- \partial_2^n u(T-\Delta t, W_{T-\Delta t})\\
= &-\frac{(\Delta t)}{n+1}g(T-\Delta t, W_{T-\Delta t}) - \frac{(\Delta t)^2}{n+2}\mathcal{L}g(T-\Delta t, W_{T-\Delta t})\\
&-\int_{T-\Delta t}^{T}\frac{(s-(T-\Delta t))^n}{(\Delta t)^n}E[\int_{T-\Delta t}^{s}\int_{T-\Delta t}^{v_1}\mathcal{L}^2 g(v_2, W_{v_2})dv_2 dv_1|\mathcal{F}_{T-\Delta t}]ds\\
&+ \partial_2^n u(T-\Delta t, W_{T-\Delta t}) + \frac{n}{n+1}(\Delta t)\mathcal{L}\partial_2^n u(T-\Delta t, W_{T-\Delta t})\\
&+ \frac{n}{2(n+2)}(\Delta t)^2\mathcal{L}^2\partial_2^n u(T-\Delta t, W_{T-\Delta t})\\
&+ \frac{n}{(\Delta t)^n}\int_{T-\Delta t}^{T}(s-(T-\Delta t))^{n-1}
\end{aligned}$$

$$E[\int_{T-\Delta t}^{s}\int_{T-\Delta t}^{v_1}\int_{T-\Delta t}^{v_2}\mathcal{L}^3\partial_2^n u(v_3, W_{v_3})dv_3dv_2dv_1|\mathcal{F}_{T-\Delta t}]ds$$
$$-\partial_2^n u(T-\Delta t, W_{T-\Delta t})$$
$$= -\frac{(\Delta t)}{n+1}g(T-\Delta t, W_{T-\Delta t}) - \frac{(\Delta t)^2}{n+2}\mathcal{L}g(T-\Delta t, W_{T-\Delta t})$$
$$-\int_{T-\Delta t}^{T}\frac{(s-(T-\Delta t))^n}{(\Delta t)^n}E[\int_{T-\Delta t}^{s}\int_{T-\Delta t}^{v_1}\mathcal{L}^2 g(v_2, W_{v_2})dv_2dv_1|\mathcal{F}_{T-\Delta t}]ds$$
$$+\frac{n}{n+1}(\Delta t)\mathcal{L}\partial_2^n u(T-\Delta t, W_{T-\Delta t})$$
$$+\frac{n}{2(n+2)}(\Delta t)^2\mathcal{L}^2\partial_2^n u(T-\Delta t, W_{T-\Delta t})$$
$$+\frac{n}{(\Delta t)^n}\int_{T-\Delta t}^{T}(s-(T-\Delta t))^{n-1}$$
$$E[\int_{T-\Delta t}^{s}\int_{T-\Delta t}^{v_1}\int_{T-\Delta t}^{v_2}\mathcal{L}^3\partial_2^n u(v_3, W_{v_3})dv_3dv_2dv_1|\mathcal{F}_{T-\Delta t}]ds.$$

By the relation $\mathcal{L}u = -f$, we have

$$I_n = -\frac{(\Delta t)}{n+1}g(T-\Delta t, W_{T-\Delta t}) - \frac{(\Delta t)^2}{n+2}\mathcal{L}g(T-\Delta t, W_{T-\Delta t})$$
$$-\int_{T-\Delta t}^{T}\frac{(s-(T-\Delta t))^n}{(\Delta t)^n}E[\int_{T-\Delta t}^{s}\int_{T-\Delta t}^{v_1}\mathcal{L}^2 g(v_2, W_{v_2})dv_2dv_1|\mathcal{F}_{T-\Delta t}]ds$$
$$-\frac{n}{n+1}(\Delta t)g(T-\Delta t, W_{T-\Delta t}) - \frac{n}{2(n+2)}(\Delta t)^2\mathcal{L}g(T-\Delta t, W_{T-\Delta t})$$
$$-\frac{n}{(\Delta t)^n}\int_{T-\Delta t}^{T}(s-(T-\Delta t))^{n-1}$$
$$E[\int_{T-\Delta t}^{s}\int_{T-\Delta t}^{v_1}\int_{T-\Delta t}^{v_2}\mathcal{L}^2 g(v_3, W_{v_3})dv_3dv_2dv_1|\mathcal{F}_{T-\Delta t}]ds$$
$$= -(\Delta t)g(T-\Delta t, W_{T-\Delta t}) - \frac{1}{2}(\Delta t)^2\mathcal{L}g(T-\Delta t, W_{T-\Delta t})$$
$$-\int_{T-\Delta t}^{T}\frac{(s-(T-\Delta t))^n}{(\Delta t)^n}E[\int_{T-\Delta t}^{s}\int_{T-\Delta t}^{v_1}\mathcal{L}^2 g(v_2, W_{v_2})dv_2dv_1|\mathcal{F}_{T-\Delta t}]ds$$
$$-\frac{n}{(\Delta t)^n}\int_{T-\Delta t}^{T}(s-(T-\Delta t))^{n-1}$$
$$E[\int_{T-\Delta t}^{s}\int_{T-\Delta t}^{v_1}\int_{T-\Delta t}^{v_2}\mathcal{L}^2 g(v_3, W_{v_3})dv_3dv_2dv_1|\mathcal{F}_{T-\Delta t}]ds.$$

We apply Itô's formula to $g(T, W_T)$ to obtain

$$
\begin{aligned}
&(\Delta t)E[g(T, W_T)|\mathcal{F}_{T-\Delta t}] \\
&= (\Delta t)g(T-\Delta t, W_{T-\Delta t}) + (\Delta t)E[\int_{T-\Delta t}^{T} \mathcal{L}g(v_1, W_{v_1})dv_1|\mathcal{F}_{T-\Delta t}].
\end{aligned}
$$

Applying Itô's formula to $\mathcal{L}g(v_1, W_{v_1})$, we see that

$$
\begin{aligned}
&(\Delta t)E[g(T, W_T)|\mathcal{F}_{T-\Delta t}] \\
&= (\Delta t)g(T-\Delta t, W_{T-\Delta t}) + (\Delta t)^2\mathcal{L}g(T-\Delta t, W_{T-\Delta t}) \\
&+ (\Delta t)E[\int_{T-\Delta t}^{T}\int_{T-\Delta t}^{v_1} \mathcal{L}^2 g(v_2, W_{v_2})dv_2dv_1|\mathcal{F}_{T-\Delta t}],
\end{aligned}
$$

and by applying Itô's formula to $\mathcal{L}^2 g(v_2, W_{v_2})$, we have

$$
\begin{aligned}
&(\Delta t)E[g(T, W_T)|\mathcal{F}_{T-\Delta t}] \\
&= (\Delta t)g(T-\Delta t, W_{T-\Delta t}) + (\Delta t)^2\mathcal{L}g(T-\Delta t, W_{T-\Delta t}) \\
&\qquad + \frac{(\Delta t)^3}{2}\mathcal{L}^2 g(T-\Delta t, W_{T-\Delta t}) \\
&\qquad + (\Delta t)E[\int_{T-\Delta t}^{T}\int_{T-\Delta t}^{v_1}\int_{T-\Delta t}^{v_2} \mathcal{L}^3 g(v_3, W_{v_3})dv_3dv_2dv_1|\mathcal{F}_{T-\Delta t}].
\end{aligned}
$$

Thus, we now obtain

$$
\begin{aligned}
&I_n + (\Delta t)E[g(T, W_T)|\mathcal{F}_{T-\Delta t}] \\
&= \frac{(\Delta t)^2}{2}\mathcal{L}g(T-\Delta t, W_{T-\Delta t}) + \frac{(\Delta t)^3}{2}\mathcal{L}^2 g(T-\Delta t, W_{T-\Delta t}) \\
&- \int_{T-\Delta t}^{T} \frac{(s-(T-\Delta t))^n}{(\Delta t)^n} \int_{T-\Delta t}^{s}\int_{T-\Delta t}^{v_1} E[\mathcal{L}^2 g(v_2, W_{v_2})|\mathcal{F}_{T-\Delta t}]dv_2dv_1ds \\
&- \frac{n}{(\Delta t)^n}\int_{T-\Delta t}^{T} (s-(T-\Delta t))^{n-1} \\
&\int_{T-\Delta t}^{s}\int_{T-\Delta t}^{v_1}\int_{T-\Delta t}^{v_2} E[\mathcal{L}^2 g(v_3, W_{v_3})|\mathcal{F}_{T-\Delta t}]dv_3dv_2dv_1ds \\
&+ (\Delta t)\int_{T-\Delta t}^{T}\int_{T-\Delta t}^{v_1}\int_{T-\Delta t}^{v_2} E[\mathcal{L}^3 g(v_3, W_{v_3})|\mathcal{F}_{T-\Delta t}]dv_3dv_2dv_1
\end{aligned}
$$

By setting

$$C_k \equiv C_k(W_{T-\Delta t}) := \sup_{s\in[T-\Delta t,T]} |E[\mathcal{L}^k g(s, W_s)|\mathcal{F}_{T-\Delta t}]|$$

for $k = 2, 3$,

$$\begin{aligned}
& I_n + (\Delta t)E[g(T, W_T)|\mathcal{F}_{T-\Delta t}] - \frac{(\Delta t)^2}{2}\mathcal{L}g(T - \Delta t, W_{T-\Delta t}) \\
& \quad - \frac{(\Delta t)^3}{2}\mathcal{L}^2 g(T - \Delta t, W_{T-\Delta t}) \\
& \quad \le \int_{T-\Delta t}^{T} \frac{(s-(T-\Delta t))^n}{(\Delta t)^n} \int_{T-\Delta t}^{s}\int_{T-\Delta t}^{v_1} C_2 dv_2 dv_1 ds \\
& \quad + \frac{1}{(\Delta t)^n}\int_{T-\Delta t}^{T} (s-(T-\Delta t))^{n-1} n \int_{T-\Delta t}^{s}\int_{T-\Delta t}^{v_1}\int_{T-\Delta t}^{v_2} C_2 dv_3 dv_2 dv_1 ds \\
& \quad + (\Delta t)\int_{T-\Delta t}^{T}\int_{T-\Delta t}^{v_1}\int_{T-\Delta t}^{v_2} C_3 dv_3 dv_2 dv_1 \\
& = \frac{C_2}{2(\Delta t)^n}\int_{T-\Delta t}^{T} (s-(T-\Delta t))^{n+2} ds + \frac{nC_2}{6(\Delta t)^n}\int_{T-\Delta t}^{T} (s-(T-\Delta t))^{n+2} ds \\
& + \frac{C_3}{6}(\Delta t)^4 \\
& = \frac{C_2}{2(n+3)}(\Delta t)^3 + \frac{nC_2}{6(n+3)}(\Delta t)^3 + \frac{C_3}{6}(\Delta t)^4. \\
& = \frac{C_2}{6}(\Delta t)^3 + \frac{C_3}{6}(\Delta t)^4.
\end{aligned}$$

Therefore, the error for $k = 2$ is given as

$$\begin{aligned}
& I_n + (\Delta t)E[g(T, W_T)|\mathcal{F}_{T-\Delta t}] \\
& \le \frac{(\Delta t)^2}{2}\mathcal{L}g(T - \Delta t, W_{T-\Delta t}) + \frac{(\Delta t)^3}{2}\mathcal{L}^2 g(T - \Delta t, W_{T-\Delta t}) \\
& + \frac{C_2}{6}(\Delta t)^3 + \frac{C_3}{6}(\Delta t)^4.
\end{aligned}$$

For $k = 3$, noting that

$$\begin{aligned}
& E[\mathcal{L}g(T, W_T)|\mathcal{F}_{T-\Delta t}] \\
& = \mathcal{L}g(T - \Delta t, W_{T-\Delta t}) + E[\int_{T-\Delta t}^{T} \mathcal{L}^2 g(v_1, W_{v_1}) dv_1|\mathcal{F}_{T-\Delta t}],
\end{aligned}$$

we see that

$$
\begin{aligned}
& I_n + (\Delta t)E[g(T, W_T)|\mathcal{F}_{T-\Delta t}] - \frac{(\Delta t)^2}{2}E[\mathcal{L}g(T, W_T)|\mathcal{F}_{T-\Delta t}] \\
& \leq \frac{(\Delta t)^3}{2}\mathcal{L}^2 g(T-\Delta t, W_{T-\Delta t}) \\
& + \frac{C_2}{6}(\Delta t)^3 + \frac{C_3}{6}(\Delta t)^4 + \frac{(\Delta t)^2}{2}\int_{T-\Delta t}^{T} C_2 dv_1 \\
& = \frac{(\Delta t)^3}{2}\mathcal{L}^2 g(T-\Delta t, W_{T-\Delta t}) + \frac{C_2}{6}(\Delta t)^3 + \frac{C_3}{6}(\Delta t)^4 + \frac{C_2}{2}(\Delta t)^3.
\end{aligned}
$$

□

3 Application

In Sect. 3.1, we recall the main result in [13], namely Theorem 2 in the end of the section. To state it, we need a series of preparations; notations, definitions, and some lemmas without proof to understand the backgrouds. In Sect. 3.2, we recall the main result in [2], that is, the non-linear discrete Clark-Ocone formula (18) in Theorem 4. In Sect. 3.3, we give the above-mentioned numerical scheme for solving (6), and explain why Theorem 1 is important for the scheme.

3.1 A Non-linear Stochastic Taylor Expansion of the Solution of (6)

We keep the notations given in Sect. 1. We start with

Lemma 1 ([13]) *For $w \in C_p^\infty([0,T]\times\mathbf{R})$, $0 \leq s \leq t \leq T$ and $n \in \mathbf{Z}_+$, let $v_t^n \in C_p^\infty([0,t]\times\mathbf{R})$ be defined by*

$$
\begin{aligned}
v_t^n(s,x) &:= \sum_{k=0}^{n}(-1)^k\frac{1}{k!}E[\mathcal{L}^k w(t, W_t)|\mathcal{F}_s^W](t-s)^k\Big|_{W_s=x} \quad (s<t), \\
v_t^n(t,x) &:= w(t,x),
\end{aligned}
\tag{13}
$$

where $\{\mathcal{F}_t^W\}$ is the natural filtration of W. Let

$$
C := \frac{2^n}{n!}\sup_{k\leq n+1, r-s\in[0,T], x\in\mathbf{R}} e^{-|x|}E[|\mathcal{L}^k w(r, W_r)||\mathcal{F}_s^W]\Big|_{W_s=x}.
$$

Then,

$$\sup_x e^{-|x|} |v_t^n(s,x) - w(s,x)| \le C(t-s)^{n+1}.$$

This lemma is obtained by applying Itô formula repeatedly, which is actually what we have done in the proof of Theorem 1. The expression of $\mathcal{L}^k w(t,x)$ for $k \in \mathbf{N}$ in (1) involves the time derivatives of w. By the relation, if we substitute u, the solution of (6) for w, we can rewrite $\mathcal{L}^k u(t,x)$ without the time derivatives by the relation $\mathcal{L}u(t,x) = -f(t,x,u(t,x),\partial_2 u(t,x))$.

Let $\mathcal{P}(f) \subset C_p^\infty([0,T]\times\mathbf{R}^3)$ for a function $f \in C_p^\infty([0,T]\times\mathbf{R}^3)$ denote the set of all polynomials over $\mathbf{R}$ in the variables

$$\partial_1^{k_1}\partial_2^{k_2}\partial_3^{k_3}\partial_4^{k_4} f(t,x,y,z), \quad (k_1,k_2,k_3,k_4) \in \mathbf{Z}_+^4,$$

or, to be precise,

$$\mathcal{P}(f) = \left\{ h\left(\partial_1^{k_1}\partial_2^{k_2}\partial_3^{k_3}\partial_4^{k_4} f(t,x,y,z); \quad (k_1,k_2,k_3,k_4) \in \mathbf{Z}_+^4\right); h \in \mathcal{P}\right\}$$

where $\mathcal{P}$ is the set of all polynomials in variables x_{k_1,k_2,k_3,k_4} for $(k_1,k_2,k_3,k_4) \in \mathbf{Z}_+^4$, and

$$h\left(\partial_1^{k_1}\partial_2^{k_2}\partial_3^{k_3}\partial_4^{k_4} f(t,x,y,z); \quad (k_1,k_2,k_3,k_4) \in \mathbf{Z}_+^4\right)$$

is the composite function of

$$h\left(x_{k_1,k_2,k_3,k_4}; \quad (k_1,k_2,k_3,k_4) \in \mathbf{Z}_+^4\right)$$

in $\mathcal{P}$ and

$$x_{k_1,k_2,k_3,k_4} = \partial_1^{k_1}\partial_2^{k_2}\partial_3^{k_3}\partial_4^{k_4} f(t,x,y,z).$$

Here we denote by $\partial_1 f$, $\partial_2 f$, $\partial_3 f$ and $\partial_4 f$ the partial derivative with respect to the first variable, the second one, the third one, and the fourth one respectively. Further, for a generic $w(t,x) \in C_p^\infty([0,T]\times\mathbf{R})$, $\mathcal{P}^w(f) \subset C_p^\infty([0,T]\times\mathbf{R})$ denotes the set of all polynomials over $\mathbf{R}$ in the functions

$$\partial_1^{k_1}\partial_2^{k_2}\partial_3^{k_3}\partial_4^{k_4} f(t,x,y,z)\big|_{y=w(t,x),z=\partial_2 w(t,x)}, \quad (k_1,k_2,k_3,k_4) \in \mathbf{Z}_+^4.$$

Let also $\mathcal{P}(v; R)$ denote the set of all polynomials over a ring R in the variables $\partial_x^k v$, $k = 1, 2, \cdots$, for a generic $v \in C_p^\infty(\mathbf{R})$.

Lemma 2 ([13]) *For $f \in C_p^\infty([0,T]\times\mathbf{R}^3)$ and $w \in C_p^\infty([0,T]\times\mathbf{R})$, we have*

$$g \in \mathcal{P}^w(f) \implies \partial_2 g \in \mathcal{P}(w(t,\cdot);\mathcal{P}^w(f))$$

and

$$g \in \mathcal{P}(w(t,\cdot);\mathcal{P}^w(f)) \implies \partial_2 g \in \mathcal{P}(w(t,\cdot);\mathcal{P}^w(f)).$$

Remark 1 We can associate the above correspondence with a nonlinear map $\tilde{\partial}_x : \mathcal{P}(w(t,\cdot);\mathcal{P}(f)) \to \mathcal{P}(w(t,\cdot);\mathcal{P}(f))$ by $\tilde{\partial}_x g_w = \partial_2 g_w$ for $g_w \in \mathcal{P}(w(t,\cdot);\mathbf{R})$ and

$$\begin{aligned}\tilde{\partial}_x g_f = a\sum_{i=1}^{m}\Big(\big(&\partial_1^{k_{i,1}}\partial_2^{k_{i,2}+1}\partial_3^{k_{i,3}}\partial_4^{k_{i,4}} f(t,x,y,z)\\ &+\partial_1^{k_{i,1}}\partial_2^{k_{i,2}}\partial_3^{k_{i,3}+1}\partial_4^{k_{i,4}} f(t,x,y,z)\partial_2 w(t,x)\\ &+\partial_1^{k_{i,1}}\partial_2^{k_{i,2}}\partial_3^{k_{i,3}}\partial_4^{k_{i,4}+1} f(t,x,y,z)\partial_2^2 w(t,x)\big)\\ &\times\prod_{\substack{j=1\\ j\neq i}}^{m}\partial_1^{k_{j,1}}\partial_2^{k_{j,2}}\partial_3^{k_{j,3}}\partial_4^{k_{j,4}} f(t,x,y,z)\Big)\end{aligned}$$

for

$$g_f = a\prod_{i=1}^{m}\partial_1^{k_{i,1}}\partial_2^{k_{i,2}}\partial_3^{k_{i,3}}\partial_4^{k_{i,4}} f(t,x,y,z),$$

and

$$\partial_2(g_f g_w) = (\partial_2 g_f)g_w + g_f(\partial_2 g_w). \tag{14}$$

From now on, we fix a $w \in C_p^\infty([0,T]\times\mathbf{R})$ satisfying

$$\mathcal{L}w(t,x) = -f(t,x,w(t,x),\partial_2 w(t,x)).$$

Lemma 3 ([13]) *We have*

$$g \in \mathcal{P}^w(f) \implies \mathcal{L}(g) \in \mathcal{P}(w(t,\cdot);\mathcal{P}^w(f)). \tag{15}$$

Remark 2 As in Remark 1, we can also associate the above correspondence a nonlinear map $\widetilde{\mathcal{L}}_0 : \mathcal{P}(f) \to \mathcal{P}(w(t,\cdot);\mathcal{P}(f))$ defined by $\widetilde{\mathcal{L}}_0 = \tilde{\partial}_t + \frac{1}{2}\tilde{\partial}_x^2$ where $\tilde{\partial}_x^2$ is the composition $\tilde{\partial}_x \circ \tilde{\partial}_x$ (restricted to $\mathcal{P}(f)$) of the map $\tilde{\partial}_x$ defined in Remark 1, and $\tilde{\partial}_t$ is defined by the relation

$$\tilde{\partial}_t g_f = a \sum_{i=1}^{m} \Big(\big(\partial_1^{k_{i,1}+1} \partial_2^{k_{i,2}} \partial_3^{k_{i,3}} \partial_4^{k_{i,4}} f(t,x,y,z)$$

$$+ \partial_1^{k_{i,1}} \partial_2^{k_{i,2}} \partial_3^{k_{i,3}+1} \partial_4^{k_{i,4}} f(t,x,y,z)(-f(t,x,y,z) - \frac{1}{2}\partial_2^2 w(t,x))$$

$$+ \partial_1^{k_{i,1}} \partial_2^{k_{i,2}} \partial_3^{k_{i,3}} \partial_4^{k_{i,4}+1} f(t,x,y,z)$$

$$\cdot (-\partial_2 f(t,x,y,z) - \partial_3 f(t,x,y,z)\partial_2 w(t,x)$$

$$- \partial_4 f(t,x,y,z)\partial_2^2 w(t,x) - \frac{1}{2}\partial_2^3 w(t,x))\big)$$

$$\times \prod_{\substack{j=1 \\ j \neq i}}^{m} \partial_1^{k_{j,1}} \partial_2^{k_{j,2}} \partial_3^{k_{j,3}} \partial_4^{k_{j,4}} f(t,x,y,z) \Big)$$

for

$$g_f = a \prod_{i=1}^{m} \partial_1^{k_{i,1}} \partial_2^{k_{i,2}} \partial_3^{k_{i,3}} \partial_4^{k_{i,4}} f(t,x,y,z).$$

Lemma 4 ([13]) *We have*

$$g \in \mathcal{P}(w(t,\cdot); \mathcal{P}^w(f)) \implies \mathcal{L}(g) \in \mathcal{P}(w(t,\cdot); \mathcal{P}^w(f)).$$

Remark 3 By defining $\widetilde{\mathcal{L}}_1 \equiv \tilde{\partial}_t + \frac{1}{2}\partial_2^2 : \mathcal{P}(w(t,\cdot), \mathbf{R}) \to \mathcal{P}(w(t,\cdot); \mathcal{P}(f))$ by

$$\tilde{\partial}_t w = -f - \frac{1}{2}\partial_2^2 w$$

and the relation

$$\tilde{\partial}_t g_w = \sum_{i=1}^{m} \partial_2^{l_i} \tilde{\partial}_t w \prod_{\substack{j=1 \\ j \neq i}}^{m} \partial_2^{l_j} w$$

implied by Leibniz's rule, we can lift $\mathcal{L}$ to $\widetilde{\mathcal{L}}[t] : \mathcal{P}(w(t,\cdot); \mathcal{P}(f)) \to \mathcal{P}(w(t,\cdot); \mathcal{P}(f))$ by

$$\widetilde{\mathcal{L}}[t](g_f g_w) = (\widetilde{\mathcal{L}}_0 g_f) g_w + g_f(\widetilde{\mathcal{L}}_1 g_w) + (\tilde{\partial}_x g_f)(\partial_2 g_w)$$

for the monomial $g = g_f g_w$ where $g_f \in \mathcal{P}(f)$ and $g_w \in \mathcal{P}(w(t,\cdot), \mathbf{R})$.

Lemma 5 ([13]) *For any $k \in \mathbf{N}$, we have*

$$\mathcal{L}^k w(t,\cdot) = (\widetilde{\mathcal{L}}^k[t])(w)(t,\cdot,y,z)\Big|_{y=w(t,\cdot),z=\partial_2 w(t,\cdot)} \in \mathcal{P}(w(t,\cdot);\mathcal{P}^w(f)).$$

As mentioned above, $\mathcal{L}^k w$ can be expanded without the time derivatives of u. Then we have the following by Lemma 1.

Lemma 6 ([13]) *Let*

$$u_T^n(s,x) = \sum_{k=0}^{n} \frac{(-1)^k}{k!} E[\widetilde{\mathcal{L}}^k[T](\Phi)(T, W_T, \Phi(W_T), \Phi'(W_T))|\mathcal{F}_s^W](T-s)^k\Big|_{W_s=x}$$

for $s < T$. *Then,*

$$\sup_x e^{-|x|}|u_T^n(s,x) - u(s,x)| = O((T-s)^{n+1}).$$

The above lemmas lead to the following

Theorem 2 ([13]) *Let*

$$u^n(t,x) \coloneqq \sum_{k=0}^{n} (-1)^k \frac{1}{k!} E[\widetilde{\mathcal{L}}^k(\Phi)(T, W_T, \Phi(W_T), \Phi'(W_T))|\mathcal{F}_t^W](T-t)^k\Big|_{W_t=x},$$

for $t < T$. *Then,*

$$\sup_x |u_T^n(t,x) - u(t,x)| = O((T-t)^{n+1}).$$

3.2 Non-linear Discrete Clark-Ocone Formla

Before stating the non-linear discrete Clark-Ocone formla, we recall the results of [1].

Let X be a square integrable differentiable function. Then, for $T > 0$, it holds that

$$X = E[X] + \int_0^T E[D_s X|\mathcal{F}_s]dW_s,$$

where D_s means the Malliavin derivative (evaluated at s). The formula plays very important role in Malliavin calculus.

A discrete-time version of Clark-Ocone formula was introduced in [1]. The dimension d is set to be one. Let $\Delta W_k = W_{k\Delta t} - W_{(k-1)\Delta t}$ for $k \in \mathbf{N}$, where $\Delta t := \frac{T}{N} (N \in \mathbf{N})$ is a fixed constant. Then, for a fixed n, the random variable

$(\Delta W_1, \cdots, \Delta W_n)$ is distributed as $N(0, \Delta t I)$. Let $\mathcal{G}_k$, $k = 1, \cdots N$, be the sigma-algebra generated by $(\Delta W_1, \cdots, \Delta W_k)$. Note that $\{\mathcal{G}_k\}_{k=0}^N$ is a filtration, and

$$L^2(\mathcal{G}_N, P) \simeq L^2(\mathbf{R}^N, \mu^N),$$

where

$$\mu^N(dx) = \frac{1}{(2\pi\Delta t)^{\frac{N}{2}}} e^{-\frac{|x|^2}{2\Delta t}} dx.$$

Theorem 3 (A Discrete Version of Clark-Ocone Formula, [1]) *For $X \in L^2(\mathcal{G}_N, P) \simeq L^2(\mathbf{R}^N, \mu^N)$, we have the following L^2-convergent series expansion:*

$$X - E[X] = \sum_{m=1}^{\infty} \sum_{l=1}^{N} \frac{(\Delta t)^{\frac{m}{2}}}{\sqrt{m!}} E[\partial_l^m X | \mathcal{G}_{l-1}] H_m \left(\frac{\Delta W_l}{\sqrt{\Delta t}} \right) \tag{16}$$

where H_m is the m-th Hermite polynomial for $m \in \mathbf{Z}_+$;

$$H_m(x) = \frac{(-1)^m}{\sqrt{m!}} e^{\frac{x^2}{2}} \frac{d^m}{dx^m} e^{-\frac{x^2}{2}} \quad (m \in \mathbf{Z}_+).$$

Here the differentiations are understood in the distribution sense.

The results in [1] are summarized as follows.

- When X^N and X are smooth, the following central limit theorem for n-th order approximation holds:

$$\begin{pmatrix} \mathrm{Err}_N(0) \\ (\Delta t)^{-\frac{1}{2}} \mathrm{Err}_N(1) \\ \vdots \\ (\Delta t)^{-\frac{n}{2}} \mathrm{Err}_N(n) \end{pmatrix} \to \begin{pmatrix} \int_0^T E[D_t X | \mathcal{G}_t] dW_t \\ \frac{1}{\sqrt{2}} \int_0^T E[D_t^2 X | \mathcal{G}_t] dB_t^1 \\ \vdots \\ \frac{1}{\sqrt{(n+1)!}} \int_0^T E[D_t^{n+1} X | \mathcal{G}_t] dB_t^n \end{pmatrix}$$

in law when $|\Delta t| \to 0$, where

$$\mathrm{Err}_N(k) := \sum_{m=k+1}^{\infty} \sum_{l=1}^{N} \frac{(\Delta t)^{\frac{m}{2}}}{\sqrt{m!}} E[\partial_l^m X | \mathcal{G}_{l-1}] H_m \left(\frac{\Delta W_l}{\sqrt{\Delta t}} \right)$$

and $(B^1, \cdots, B^n) = (B_t^1, \cdots, B_t^n)_{0 \le t \le T}$ is Brownian motion that is independent of $W = (W_t)_{0 \le t \le T}$.

- If X is only fractionally differentiable for $s \in (0, 1)$, then the mean squared error of the first order approximation is $O(N^{-\frac{s}{2}})$.

Remark 4 In [12], it is shown that the mean squared error of n-th approximation is $O(N^{-\frac{sn}{2}})$, if X is only fractionally differentiable for $s \in (0, 1)$. In the context of mathematical finance, the result indicates that gamma hedge and higher-order hedge are useful for an option that is difficult to be hedged as discontinuity of the corresponding payoff function, for instance, binary option.

Let us consider a discrete-time analog of the Eq. (1) as follows. Here we work on any $d \ge 1$.

Definition 1 For a given adapted process X^N and $g_T \in L^2(\mathcal{G}_N, P)$, if there exist adapted process Y^N and Z^N such that

$$\begin{aligned} Y_t^N = g_T + \sum_{j=t}^{N-1} f_j(X_j^N, Y_j^N, Z_j^N)\Delta t \\ + \sum_{j=t}^{N} Z_j^N \cdot \Delta W_j + M_t, \end{aligned} \tag{17}$$

where M is a martingale orthogonal to $\sum_{j=t}^N Z_j^N \cdot \Delta W_j$, then we call the pair of adapted process Y^N and the d-dimensional predictable process Z^N a solution to the discrete BSDE (17).

Theorem 4 (Non-Linear Discrete Clark-Ocone Formula [2]) *Let $y \mapsto y - f(t, x, y, z)\Delta t : \mathbf{R} \to \mathbf{R}$ be a surjective. Then, there is a solution to the discrete BSDE* (17)*, and the solution is given by the following recursive equation: Let $Y_N = g_T$ and $Z_N^N = E[\partial_N g_T | \mathcal{G}_{N-1}]$, and for $j = N-1, N-2, \cdots, 1$,*

$$Y_{j-1} - f((j-1)\Delta t, X_{j-1}, Y_{j-1}, E[\partial_j Y_j | \mathcal{G}_{j-1}])\Delta t = E[Y_j | \mathcal{G}_{j-1}], \tag{18}$$

and

$$Z_{j-1} = E[\partial_j Y_j | \mathcal{G}_{j-1}],$$

where we write $\partial_j \equiv (\partial_{\Delta W_j^1}, \cdots, \partial_{\Delta W_j^d})$. Moreover, we have the following non-linear Clark-Ocone formula; for $j = N, N-1, \cdots, 1$,

$$Y_{j-1} - Y_j - f((j-1)\Delta t, X_{j-1}, Y_{j-1}, E[\partial_j Y_j | \mathcal{G}_{j-1}])\Delta t$$
$$= \sum_{m=1}^{\infty} \sum_{k_1+\cdots+k_d=m} \frac{(\Delta t)^{\frac{m}{2}}}{\sqrt{m!}} E[\partial^{k_1}_{\Delta W^1_j} \cdots \partial^{k_d}_{\Delta W^d_j} Y_j | \mathcal{G}_{j-1}] \times \prod_{i=1}^{d} H_{k_i}(\frac{\Delta W^i_j}{\sqrt{\Delta t}}). \tag{19}$$

Moreover, if $y \mapsto f(t, x, y, z)\Delta t - y : \mathbf{R} \to \mathbf{R}$ *be a bijection, the solution is unique.*

3.3 A Higher-Order Scheme for Solving the Semi-linear PDE (6)

Before stating our new scheme/alrorithm, we recall three of preceding ones, which inspired our scheme very much.

The method proposed by W. E, J. Han and A. Jentzen [8] is summarized as follows. The approximation $u(t_i, \cdot)$ of the solution u of (6) is given by

$$u(t_{i+1}, X_{i+1}) \approx F(t_i, W_{t_i}, u(t_i, X_{t_i}), \partial_2 u(t_i, W_{t_i}), \Delta t_i, \Delta W_{t_i})$$

with

$$F(t, x, y, z, h, \Delta) := y - f(t, x, y, z)h + z^{\mathsf{T}}\Delta, \tag{20}$$

where the function $x \in \mathbf{R}^d \mapsto \partial_2 u(t_i, x)$ for each i, as well as $x \mapsto u(t_0, x)$ is approximated by a multi-layer neural network parameterized by θ_i $x \in \mathbf{R}^d \mapsto \mathcal{Z}_i(x; \theta_i)(i = 1, \cdots, N-1)$. Then, an estimator $\mathcal{U}_i$ of $u(t_i, \cdot)$ for each i can be obtained, forward-recursively in i, as

$$\mathcal{U}_{i+1} = F(t_i, W_{t_i}, \mathcal{U}_i, \mathcal{Z}_i(X_{t_i}; \theta_i), \Delta t_i, \Delta W_{t_i}),$$

$i = 1, \cdots, N-1$. The parameter θ is obtained via the stochastic gradient descent method (SGD for short) as the minimizer of

$$\theta \mapsto E|g(X_{t_N}) - \mathcal{U}_N(\theta))|^2.$$

Second, we recall the algorithm proposed by Hure et al. [9].

- Initialize from an estimation $\hat{\mathcal{U}}_N$ of $u(t_N, \cdot)$ with $\hat{\mathcal{U}}_N = g$.

- For $i = N-1, \cdots, 0$, given $\hat{\mathcal{U}}_{i+1}$, use a deep neural network $\mathcal{U}_i(\cdot;\theta)$ and compute (by SGD) the minimizer of the expected quadratic loss function:

$$\begin{aligned}&\hat{L}_i(\theta)\\&:= E|\hat{\mathcal{U}}_{i+1}(W_{t_{i+1}})\\&- F(t_i, W_{t_i}, \mathcal{U}_i(W_{t_i};\theta), \widehat{\partial_2 \mathcal{U}_i}(W_{t_i};\theta),\\&\qquad\qquad\qquad\qquad \Delta t_i, \Delta W_{t_i})|^2,\\&\theta_i{}^* \in \mathrm{argmin}_{\theta\in\mathbf{R}^{N_m}} \hat{L}_i(\theta)\end{aligned} \tag{21}$$

  where $\widehat{\nabla_x \mathcal{U}_i}(W_{t_i};\theta)$ is either

  – a neural network function (DBDP1 in [9]), or
  – a numerical differentiation of $\mathcal{U}_i(\,\cdot\,;\theta)$ (DBDP2 in [9]).

  Then, update : $\hat{\mathcal{U}}_i = \mathcal{U}_i(\,\cdot\,;\theta_i{}^*)$.

  Third, let us recall the algorith proposed in [2], where F in (20) is replaced with

$$\begin{aligned}&F_M(t, x, y, z_1, \cdots, z_M, h, \Delta_1, \cdots, \Delta_M)\\&:= y - f(t, x, y, z_1)h + \sum_{m=1}^{M} h^{\frac{m}{2}} z_m \cdot \Delta_m,\end{aligned} \tag{22}$$

where z_m and $\Delta_m \in (\mathrm{R}^d)^{m\otimes}$, $m = 1, \cdots, M$, and

$$\begin{aligned}&\mathcal{U}(t_{i+1}, W_{t_{i+1}})\\&= F_M\Big(t_i, W_{t_i}, \mathcal{U}(t_i, W_{t_i};\theta),\\&(\widehat{\partial_2^m}\mathcal{U}(t_i, W_{t_i};\theta))_{m=1}^M, \Delta t_i, \left(H_m(\frac{\Delta W}{\sqrt{\Delta t}}) \times \frac{1}{\sqrt{m!}}\right)_{m=1}^M\Big).\end{aligned} \tag{23}$$

Here we set, for $(x^1, \cdots, x^d) \in \mathbf{R}^d$ and $(i_1, \cdots, i_m) \in \{1, \cdots, d\}^m$,

$$(\mathrm{H}_m(x^1, \cdots, x^d))_{i_1,\cdots,i_m} = \prod_{k=1}^{d} H_{\sharp\{l:i_l=k\}}(x^k).$$

As in the algorithms of [9] explained above, in each step we approximate $\mathcal{U}(t_i, W_{t_i};\theta)$ by a neural network function and $\widehat{\nabla_x^{\otimes m}\mathcal{U}}$ by either neural network functions or numerical approximation of derivatives, and find minimizer θ by SGD. The algorithm is implied by the M-th truncation of the formula (19) in Theorem 4.

Now we introduce our new scheme. We replace F_M of (22) with

$$F_{0,M}(t, x, y, z_1, \cdots, z_M, h, \Delta_1, \cdots, \Delta_M)$$
$$:= y + \sum_{m=1}^{M} h^{\frac{m}{2}} z_m \cdot \Delta_m,$$

and consider the error function

$$\Bigg| F_{0,M}\Big(t_i, W_{t_i}, \mathcal{U}(t_i, W_{t_i}; \theta), (\widehat{\partial_2^m}\mathcal{U}(t_i, W_{t_i}; \theta))_{m=1}^M, \Delta t_i, \Big(H_m(\frac{\Delta W}{\sqrt{\Delta t}}) \times \frac{1}{\sqrt{m!}}\Big)_{m=1}^M\Big) - \sum_{k=0}^{n} (-1)^k \frac{h^k}{k!} \tilde{\mathcal{L}}^k (u(t_{i+1}, \cdot))(T, \cdot, u(t_{i+1}, \cdot), \partial_2 u(t_{i+1}, \cdot)) \Bigg|^2, \tag{24}$$

where we take M sufficiently large so that the minimizers $(\widehat{\partial_2^m}\mathcal{U}(t_i, W_{t_i}; \theta^*))_{m=1}^M$ of the error function (24) can be used for $u(t_i, \cdot)$ in the next step.

We note that the minimizer satisfies

$$\widehat{\partial_2^m}\mathcal{U}(t_i, W_{t_i}; \theta^*)$$
$$= E[\partial_2^m \sum_{k=0}^{n} (-1)^k \frac{h^k}{k!} \tilde{\mathcal{L}}^k (u(t_{i+1}, \cdot))(T, \cdot, u(t_{i+1}, \cdot), \partial_2 u(t_{i+1}, \cdot)) | \mathcal{F}_{t_i}],$$
$$m = 1, 2, \cdots, M,$$

by virtue of the discrete Clark-Ocone formula (16). or the non-linear version (19). Thus, Theorem 1 ensures the scheme is of order 2 or 3, respectively, though we postpone the rigorous proof to the paper in preparation.

4 Concluding Remarks

We have postponed

- describe the scheme for dealing with the general case (2),
- the error estimates of the prolongation for general k, and
- the rigorous proof that the scheme really works as a higher-order scheme; an error estimate should be established.

We acknowledge the discussions and the suggestions with/from Prof. Jirô Akahori.

References

1. Akahori, J., Amaba, T., Okuma, K.: A discrete-time Clark-Ocone formula and its application to an error analysis. J. Theor. Probab. **30**, 932–960 (2017)
2. Akahori, J., Furuichi, Y., Okuma, K.: Higher-order deep solver of non-linear PDEs implied by a non-linear discrete Clark-Ocone formula. JSIAM Lett. **30**, 9–12 (2022)
3. Bouchard, B., Touzi, N.: Discrete-time approximation and Monte-Carlo simulation of backward stochastic differential equations. Stoch. Process. Appl. **111**, 175–206 (2004)
4. Chassagneux, J.F., Crisan, D.: Runge–Kutta schemes for backward stochastic differential equations. Ann. Appl. Probab. **24**, 679–720 (2014)
5. Chassagneux, J.F., Trillos, C.A.G.: Cubature method to solve BSDEs: error expansion and complexity control. Math. Comput. **89**, 1895–1932 (2020)
6. Chessari, J., Kawai, R., Shinozaki, Y., Yamada, T.: Numerical methods for backward stochastic differential equations: a survey. Probab. Surv. **20**, 486–567 (2023)
7. Crisan, D., Manolarakis, K.: Second order discretization of backward SDEs and simulation with the cubature method. Ann. Appl. Probab. **24**, 652–678 (2014)
8. Weinan, E., Han, J., Jentzen, A.: Deep learning-based numerical methods for high-dimensional parabolic partial differential equations and backward stochastic differential equations. Commun. Math. Stat. **5**, 349–380 (2017)
9. Hure, C., Pham, H., Warin, X.: Deep backward schemes for high-dimensional nonlinear PDEs. Math. Comput. **89**, 1547–1579 (2020)
10. Ma, J., Yong, J.M.: Forward-Backward Stochastic Differential Equations and their Applications. Springer, Berlin (2007)
11. Naito, R., Yamada, T.: A second-order discretization for forward-backward SDEs using local approximations with Malliavin calculus. Monte Carlo Methods Appl. **25**, 341–361 (2019)
12. Nishimura, T., Yasutomi, K., Yuasa, T.: Higher order error estimate of the discrete-time Clark–Ocone formula. J. Theor. Probab. **35**, 2518–2539 (2022)
13. Okuma, K.: An asymptotic expansion of the solution of a semi-linear partial differential equation implied by a nonlinear Feynman-Kac formula. Int. J. Math. Industry 16, 2450002 (2024). https://doi.org/10.1142/S2661335224500023